012221775

D1758945

Please re
below. A
Books m:
another r
Recall, b
5412.

DUE FOR RETURN

11 OCT 1996

DUE TO RETURN

7 JAN 1997

CANCELLED

For condition

SOILS IN THE
URBAN ENVIRONMENT

Soils in the
Urban Environment

EDITED BY PETER BULLOCK AND
PETER J. GREGORY ON BEHALF OF
THE BRITISH SOCIETY OF SOIL SCIENCE
AND THE NATURE CONSERVANCY COUNCIL

OXFORD
BLACKWELL SCIENTIFIC PUBLICATIONS
LONDON EDINBURGH BOSTON
MELBOURNE PARIS BERLIN VIENNA

© 1991 by
Blackwell Scientific Publications
Editorial offices:
Osney Mead, Oxford OX2 0EL
25 John Street, London WC1N 2BL
23 Ainslie Place, Edinburgh EH3 6AJ
3 Cambridge Center, Cambridge
   Massachusetts 02142, USA
54 University Street, Carlton
   Victoria 3053, Australia

Other Editorial Offices:
Arnette SA
2, rue Casimir-Delavigne
75006 Paris
France

Blackwell Wissenschaft
Meinekestrasse 4
D-1000 Berlin 15
Germany

Blackwell MZV
Feldgasse 13
A-1238 Wien
Austria

First published 1991

Set by Excel Typesetters, Hong Kong
Printed and bound in Great Britain by
the University Press, Cambridge

DISTRIBUTORS

   Marston Book Services Ltd
   PO Box 87
   Oxford OX2 0DT
   (*Orders:* Tel: 0865 791155
              Fax: 0865 791927
              Telex: 837515)

USA
   Blackwell Scientific Publications, Inc.
   3 Cambridge Center
   Cambridge, MA 02142
   (*Orders:* Tel: 800 759-6102)

Canada
   Oxford University Press
   70 Wynford Drive
   Don Mills
   Ontario M3C 1J9
   (*Orders:* Tel: 416 2941)

Australia
   Blackwell Scientific Publications
   (Australia) Pty Ltd
   54 University Street
   Carlton, Victoria 3053
   (*Orders:* Tel: 03 347-0300)

British Library
Cataloguing in Publication Data

Soils in the Urban environment.
   1. Urban Regions. Soils
   I. Bullock, P. (Peter), *1937—*   II. British Society of
Soil Science   III. Nature Conservancy Council
631.481732

   ISBN 0-632-02988-9

Library of Congress
Cataloging in Publication Data

Soils in the urban environment/edited by Peter Bullock
   and Peter J. Gregory on behalf of the British Society of
   Soil Science and Nature Conservancy Council.
      p.   cm.
   Includes index.
   ISBN 0-632-02988-9
      1. Urban soils.   I. Bullock, Peter.
   II. Gregory, P. J.   III. British Society of Soil Science.
   IV. Nature Conservancy Council (Great Britain)
   S592.17.U73S65   1991
   631.4′1′091732—dc20

Contents

List of Contributors

H.J. ASH, *5 Dearnford Avenue, Bromborough, Wirrall, Merseyside L62 6DX*

E.M. BRIDGES, *Department of Geography, University of Wales, Swansea SA2 8PP*

P. BULLOCK, *Soil Survey and Land Research Centre, Silsoe, Beds MK45 4DT*

P.J. GREGORY, *Department of Soil Science, The University, Reading RG1 5AQ*

J.A. HARRIS, *Environment and Industry Research Unit, Department of Biosciences, Polytechnic of East London, London E15 4LZ*

J.M. HOLLIS, *Soil Survey and Land Research Centre, Silsoe, Beds MK45 4DT*

C.E. MULLINS, *Department of Plant and Soil Science, University of Aberdeen, Aberdeen AB9 2UE*

I.D. PULFORD, *Department of Agricultural, Food and Environmental Chemistry, University of Glasgow, Glasgow G12 8QQ*

D.L. RIMMER, *Department of Agricultural and Environmental Science, The University, Newcastle upon Tyne NE1 7RU*

I. THORNTON, *Environmental Geochemistry Research, I.C.C.E.T., Imperial College, Royal School of Mines, Prince Consort Road, London SW7 2BP*

Preface

Urban areas contain a very wide variety of open spaces including gardens, playing fields, waste lands, spoil heaps, railway and canal embankments and islands of natural or semi-natural land. For many years the tendency was to allow much of this open space not in allotments or playing fields to evolve with a minimum of management. Very little was discovered about the nature of the soils or the problems or potential inherent in them. This situation is now changing rapidly and there is increasing interest in the degree and location of contaminated land in urban areas, and in planning an attractive environment for urban dwellers and visitors.

The recent House of Commons Environment Committee First Report on Contaminated Land has highlighted many of the problems relating to this area. These include identifying the location of contaminated land, monitoring it and recognizing the implications of it, particularly for human health. These problems clearly need addressing and call for a much better knowledge base of urban soils.

Already there are ecologists active in urban areas, seeking to create new habitats and preserve valuable existing ones. A Research Advisory Group on Urban Ecology established by the Ecological Parks Trust with funding from the Nature Conservancy Council identified in their published report (*Research in Urban Ecology* by Ian Barrett, Trust for Urban Ecology, 1987) a major lack of knowledge of the development, properties and potential of urban soils. It was recognized that such information is fundamental to the planned development of open spaces if expensive mistakes are to be avoided.

The Nature Conservancy Council recognized the need for sound information about the soils of urban areas and in September 1988 set up jointly with the British Society of Soil Science a national meeting to review the state-of-the-art of knowledge of soils in urban areas. This book represents the proceedings of that meeting. It is hoped that it will provide a basis from which to plan future research and development programmes.

In editing the book on behalf of the British Society of Soil Science and the Nature Conservancy Council, the editors wish to acknowledge the support of the Nature Conservancy Council, and in particular that of G.M.A. Barker, the Urban Development Officer, and Dr L.A. Batten of the Chief Scientist Directorate.

Peter Bullock and Peter J. Gregory
1990

1 Soils: A neglected resource in urban areas

P. BULLOCK AND P.J. GREGORY

Introduction

Soils are responsible for a variety of functions, the most important of which are as a medium for plant growth, a foundation for buildings and a source and sink for water and pollutants (Table 1.1). After several centuries in which the soil cover of the UK has sustained the activities of the population without severe problems, the last 40 years have seen problems of varying severity begin to appear. These include contamination, erosion, acidification and compaction. These problems have begun to focus attention on the importance of the soil cover, the need for a better understanding of it and the need for its protection.

Although most attention has been given to soils of rural areas because of their importance in food production, there are significant areas of open space in urban areas. According to Best (1981), the urban area of Great Britain is about 1.7 million hectares. Of this, roughly 12% is classed as open space. This includes allotments, parkland and derelict land. Some of the soils of urban areas, particularly those of the allotments, have been well maintained, but others have been damaged physically, chemically and/or biologically. The main problems associated with urban soils are contamination, compaction, poor drainage and the stone content.

Increasingly, local councils are being encouraged to provide an attractive environment for urban dwellers and visitors but are restricted by a general lack of understanding and information about the soils for which they are responsible. As a result, planning land use and management in urban areas has largely been on the basis of a 'try it and see' approach rather than one based on a sound understanding of the medium in which the plants are to grow. It is little wonder, therefore, that land use management in urban areas is marked by some notable failures. For example, according to Professor A.D. Bradshaw (*New Scientist*, November 6th 1986), of the ten million trees planted annually about half die in the first five years, thereby wasting some £10 million of public money. Not all these trees are in the urban environment, and soil is not the only factor involved, but there is little doubt that the success rate in urban areas could be much improved by identifying the soil-related problems and matching species to soil conditions.

Urban areas can include significant amounts of contaminated land. The recent House of Commons Environment Committee First Report on *Contaminated Land* (1990) draws attention to the lack of information available about the location of contaminated soils and the nature of the contamination. Such sites have implications for future land use as well as for human health, highlighting the need for better information about the soils of urban areas.

This book is intended to be a review of the state-of-the-art of the science of soils of urban areas. It will be apparent from some of the chapters that there are considerable gaps in our knowledge. We hope that this will be a stimulus for improving the knowledge of the soils, their properties, and their uses thereby creating a better environment and facilitating a better use of public money.

A framework for information transfer

With the gradually increasing interest in planning the environment of urban areas, it is essential that a framework be developed on which to base research and management, and transfer experience. Just as Linnaeus established a classification system for plants which forms a basis for their identification, description and information transfer, so there is a need for an equivalent one for soils.

Classification systems for soils have been developed, but with special reference to the natural and semi-natural soils of rural areas. Two international systems now exist and most countries have their own national system. There are systems for England and Wales and for Scotland, but both are relatively weak on man-made soils, the group most relevant to the urban environment.

J.M. Hollis, in his review of existing classification

Table 1.1 Soil functions

Elemental recycling
   Nutrient absorption
   Nutrient storage
   Nutrient supply

Plant growth medium
   Food crops
   Timber crops
   Energy crops
   Other commercial exploitable plants
   Plants of natural habitats
   Plants for amenity sites

Water cycling
   Absorption
   Storage
   Supply
   Amelioration

Anchorage
   Plants

Substrate/habitat for soil fauna/flora
   Macro- and meso-fauna
   Micro-fauna and -flora

Maintenance of food supply for ground-feeders, e.g. birds

Contamination
   Source of pollutants
   Sink for pollutants

Foundations
   Low-level buildings
   Roads
   Reservoirs/ponds

systems in Chapter 2, describes classification systems for urban soils developed in a number of countries, e.g. Germany, USA, Yugoslavia, but concludes that they contain major omissions and inconsistencies. In particular, precise differentiating criteria to define the soils have not been developed. A classification system relevant to UK conditions needs to be developed incorporating biological, chemical and physical criteria.

Rubble or soils

The soils of urban areas represent a wide spectrum in terms of composition and degree of development. There are areas which appear to lack any soil cover, e.g. brick rubble, demolition sites, and may carry no vegetation. Yet, if not heavily contaminated, such areas rapidly develop a plant cover and soil development begins. At the other end of the spectrum are the well-developed soils of small enclaves of undisturbed land, comparable to those of the rural sector. All degrees of soil development between these two ends of the spectrum can be found in urban areas.

The range of materials which comprise the soils of urban areas is much broader than that of rural areas. E.M. Bridges demonstrates in Chapter 3 that these materials can be derived from a number of very different origins. In addition to the inherited sand, silt, clay and organic matter, there are often important additions of materials, e.g. debris from building sites, buried waste on disposal sites, pulverized fuel ash from power stations, residues from metalliferous industries. The full list is a long one.

Most soil research has concentrated on the soils of the rural sector and many of the processes in these soils are understood. The soils of urban areas bring together materials of very different composition and origin, and the way in which such materials are likely to interact with each other is poorly understood. The production of methane gas on waste disposal sites is an example of one such interaction.

There have been relatively few detailed systematic studies of the composition of soils in urban areas, most studies to date focusing on a particular topic such as contamination. There is an urgent need for more information on the composition of urban soils. It is only then that a better understanding can be achieved of the interactions between the main components and the implications of these for land use and health and safety.

Contaminated land

Few restraints have been put on industry previously to prevent contamination of the soil beneath and around industrial sites. The industrial revolution and subsequent decades have bequeathed a variety of contaminated sites to urban areas. The situation is further complicated by the fact that many of the industrial concerns have now disappeared or changed, and there is no clear picture of the location, extent and nature of the contaminated sites in the UK. The recent House of Commons Environment Select Committee Report on *Contaminated Land* (1990) focuses on this and several other key issues.

Some contaminated sites are well known and steps are being taken to protect the public at them. These mainly relate to heavy metal contamination. In Chapter 4, I. Thornton concludes that apart from heavy metals there is 'little systematic information on the extent and degree of urban soil contamination with pesticides, herbicides, hydrocarbons, fertilizers, asbestos, etc.'.

Chapter 4 also concludes that more research is required to understand the dynamics and mobility of contaminants in these soils and the factors affecting their availability to plants and the food chains of animals and humans. An inventory needs to be made of all contaminated land and the nature of the contaminants present, and research and technology developed to make the sites safe where necessary.

Relationships between soils and vegetation

The relationships between soils and vegetation have been poorly researched, which is surprising in view of the long history of botany. This neglect stems at least in part from the fact that ecologists have been more interested in taxonomy and the nature of what grows above the soil rather than developing an understanding of the medium in which the plants grow. To a large extent the soil has been treated as a black box.

It is hardly surprising, therefore, that there is rather little detailed information about soil–vegetation interaction in urban areas. In Chapter 9, H.J. Ash develops relationships between vegetation in urban areas and broad soil conditions in so far as they are known. It raises the fascinating question of 'Why is a particular habitat where it is?' Soil type is an important factor but there are other factors such as seed banks and seed dispersal. To create habitats in the urban environment, and this is likely to be an ongoing challenge, the role of soils and their relationship to other factors governing habitats need to be researched. In addition to creating new habitats,

it is sometimes desirable physically to move an existing habitat, for example one threatened by building development, to a new location. For this to be successful it is essential to obtain a good match in terms of soil type between the old and the new location. The science behind such transfer is only just developing. The urban environment provides an exciting and challenging arena for these studies.

The ability of soils to sustain a particular habitat once it is established is governed to an important extent, in the absence of fertilizer applications, by nutrient supply. Some nutrients such as nitrogen, potassium and phosphorus are required in relatively large amounts, whereas others, the micronutrients like boron, zinc and copper, need be available only in very small amounts. The ability of the soil to establish a nutrient cycling regime is critical to the maintenance of most types of vegetation. I.D. Pulford reviews this important area in Chapter 7 and expresses his concern about the paucity of information on nutrient availability in urban soils. The role of organic matter and the fascinating balance between nutrient availability and non-availability in habitat creation and sustainability in urban soils are areas needing further research.

Physical properties of urban soils

One of the major problems afflicting urban soils is their poor physical structure and this is one of the main causes of poor establishment of vegetation. This damage relates in part to the composition of the soil but is accentuated by the treatment to which the soil has been subjected. New developments inevitably affect the soils around the sites, landscaping changes the thickness and topography of the soil cover, and use of heavy equipment, particularly when the soil is wet, all create physical problems. In Chapter 6, C.E. Mullins discusses the principles behind the physical behaviour of soils and describes case studies relevant to the urban situation. A good example of the way in which these principles are an aid to solving problems is in relation to playing fields. Maintaining sports fields in such a manner that playability is at a maximum while at the same time keeping damage and further problems to a minimum has concerned local councils for many years. Yet as C.E. Mullins describes, if the principles behind the problem are understood, a satisfactory solution to the problem can often be found. In the case of playing fields this involves an understanding of soil drainage, the factors affecting playability, e.g. load-bearing capacity, and the requirements for sustained growth of grass.

Much of the soil in urban areas has been moved or disturbed at some time in its history. The major effects of movement of the soil are compaction and loss of structure. Little research has been carried out on the effects of movement, storage and handling of soils in the urban environment, but as D.L. Rimmer indicates in Chapter 5, substantial research has been done in connection with land restoration schemes associated with mineral extraction, particularly coal, gravel and rock. He discusses the application of this research to soils in the urban environment.

Biology of soils in urban areas

Soil biology has been a neglected subject in the UK in recent years. Little is known about the fauna and microflora of urban soils yet they play an important part in the decomposition of organic matter, nutrient cycling, formation of soil structure, concentration of pollutants and the functioning of ecosystems.

In Chapter 8, J.A. Harris reviews the state of scientific knowledge in this area. Compared to the soils of rural areas, those in the urban environment contain fewer organisms, less biomass and more restricted species diversity, reflecting the generally greater disturbance to which urban soils have been subjected. Chapter 8 also draws attention to the health risk from some soils in heavily used amenity areas, an aspect that has received little attention in the UK previously.

Soil fauna and microflora have a role to play as indicators of soil quality. It is important that methods for assessing the quality of soils are developed and the sensitivity of soil fauna and microflora to soil conditions make them particularly suitable for this purpose.

Future research needs

The book attempts to summarize the state of scientific knowledge about the soils of the urban environment for those concerned with land-use planning and management. It demonstrates clearly the disappointing information base which exists for the urban areas. Several of the authors have found it necessary to draw on the research that has been carried out on soils of rural areas. This is an important research base, much of which is applicable to urban areas. However, there are important differences between soils of urban areas and those of rural areas, not least that they are often composed of rather different materials. There is a need for research to be directed to the specific problems of the soils in urban areas and the following chapters identify some of the principal topics needing attention.

References

Best, R.H. (1981). *Land Use and Living Space*, 197pp. Methuen.

House of Commons. (1990). Environment Committee. First Report: *Contaminated Land*, Vol. 1.

2 The classification of soils in urban areas

J.M. HOLLIS

Introduction

Recent work (Craul, 1985; Barrett, 1987) has emphasized the differences that can occur between soils in the urban environment and those in a more natural, rural situation. Barrett (1987) documents the need for more information about many aspects of soils in urban areas but stresses the lack of a clear framework from which to identify gaps in research or judge the validity and importance of past and future projects. The purposes of this paper are to propose some basic principles upon which any classification of urban soils should be based if it is to provide a framework for research and technology transfer, and to review existing soil classification systems and their usefulness in the context of soils in urban areas.

Basic principles for the classification of soils in urban areas

THE NATURE OF SOIL

Soil means different things to different people. To the farmer or gardener, it is the upper few centimetres of material that he cultivates. To the engineer or geologist it is the 'overburden' or upper few metres of unconsolidated material at the surface of the Earth's crust. Before we can consider how to classify soils in urban areas, we must first decide exactly what we are classifying. Craul (1985) considers that Bockheim (1974) gives 'an appropriate and useful definition' of urban soil as 'a soil material having a non-agricultural, man-made surface layer more than 50 cm thick, that has been produced by mixing, filling, or by contamination of land surface in urban and suburban areas'. He further states 'the inference is that the soil has been at least partially disturbed in some part of the profile . . .'. However, some soils in urban areas are likely to be relatively undisturbed, or disturbed only to a shallow depth, and if these are to be included in any classification, a broader definition is necessary. Such a broad definition of soil is usually given as 'the unconsolidated mineral or organic material at the Earth's surface that is capable of supporting plant growth' (e.g. Avery, 1980; Bridges, 1982). This definition includes a whole range of materials likely to be found in an urban context, from loose, unconsolidated, man-made material unmodified by soil-forming processes, to undisturbed 'naturally developed' soils. However, as has been pointed out by Fanning *et al.* (1978) it excludes materials which in their present state are too toxic to support plants, but have the *potential* to do so, if their toxicity problems are corrected. Such materials are likely to be present in the urban environment and would need to be included in any useful classification. The definition of soil therefore needs to be modified to: *Any unconsolidated mineral or organic material at the Earth's surface that has the potential to support plant growth.*

THE NATURE AND PURPOSES OF SOIL CLASSIFICATION

The essential purpose of soil classification is to organize existing knowledge so that the properties and relationships of different kinds of soils can be systematically recalled and communicated (Cline, 1949). This is a simple statement that conceals numerous problems. Firstly, soils do not exist as discrete individuals like plants or animals, so that the unit of soil classification must be defined before any classification is developed. Secondly, there is no single *correct* way of classifying soils. The way in which soil properties and relationships are organized to distinguish classes depends upon the purpose for which the classification is created. In the case of soils, the purposes can vary greatly and many different types of classification are possible. This fundamental subject has been extensively reviewed by De Bakker (1970) and Clayden (1982) but a brief summary is necessary before we can consider the basis of any classification of urban soils.

Two main types of classification scheme are in

current use. Most soil classifications use a *hierarchical system* with a framework resembling a family tree, in which successively lower categories have more classes than the one above (Fig. 2.1). At each categorical level, properties are chosen to define mutually exclusive classes and this involves ranking the criteria to define successively lower categories. Alternatively, *coordinate systems* of classification (Avery, 1968) do not use ranked hierarchies of differentiating criteria, but comprise two or more specific classifications, each based on one set of attributes, that are superimposed to create a two- or multi-dimensional matrix of classes (Fig. 2.2). Both of these systems employ what has been called the 'taxonomic chop' (Butler, 1980) to define classes by choosing limiting values of a particular property. Thus two soil classes could be defined according to their clay content, one with 0–20% clay and the other with more than 20% clay. The class limit imposed at 20% clay is a taxonomic chop.

Both types of classification can be developed using statistically-based, numerical techniques. In a hierarchical system these are used to suggest ways of minimizing within-class differences and maximizing between-class differences, whereas in a coordinate system they can be used more fundamentally to create multidimensional classes within which variation is minimal (Webster, 1975, 1977).

Hierarchical and coordinate systems each have their advantages and disadvantages, but the choice of which system is best and whether numerical classification methods are useful for its development, depends on the purpose for which the classification is being developed. It has been argued that soil classification systems are used for one of two basic purposes, although there is every gradation between the two and surprisingly a definitive statement about their purpose is often lacking (De Bakker, 1970). The basic purposes are met by two main schemes:

1 Theoretical or scientific schemes which place emphasis on soil genesis and the relationships between classes; they make use of as many known properties of the soils as possible, without a single or applied objective and are also known as 'natural classifications'.

2 Practical schemes which are designed for application to a specific purpose or number of purposes and are often devised to serve the needs of soil survey (e.g. De Bakker & Schelling, 1966; Soil Survey Staff, 1975; Avery, 1980).

If a theoretical scheme is desired, then it is probably best to use numerical methods to construct a coordinate classification which, by means of an extensive, statistically-based, sampling scheme, would group the soil population into a number of classes using taxonomic chops based on whatever order is present within the range of properties chosen. The advantage of this is that it groups soils according to their 'natural clustering' without imposing any artificial bias. The disadvantage is that it can become very complicated and unwieldy, usually needing a

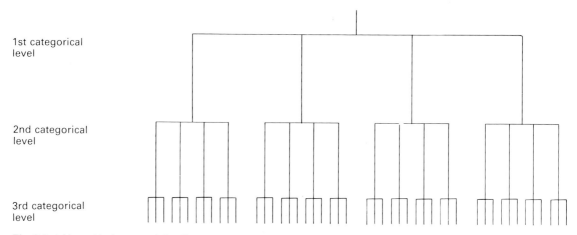

1st categorical level

2nd categorical level

3rd categorical level

Fig. 2.1 A hierarchical system of classification.

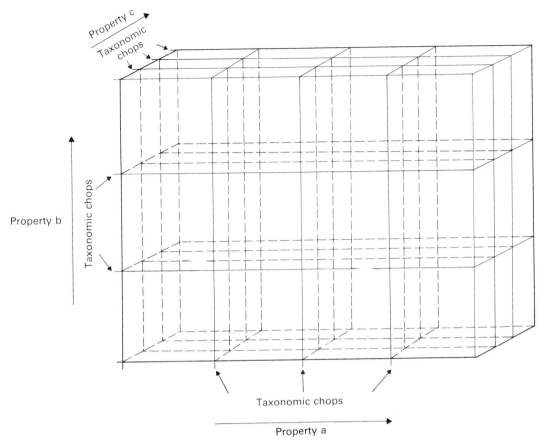

Fig. 2.2 A coordinate system of classification.

computer to manipulate and retrieve class data. Furthermore, current knowledge suggests that most soil properties are continuously variable with no natural clustering, although there may be local clustering within limited areas. This contrasts with populations of higher plants and animals which have inherent nested clustering, where individuals forming a group have many shared attributes not found in individuals from other groups. In principle, therefore, the choice of taxonomic chops within any soil classification system is arbitrary despite the fact that random sampling of the soil population may suggest otherwise.

If, on the other hand, a practical classification is desired, then it is probably best to use either a hierarchical or a coordinate system based on predetermined taxonomic chops, each of which has some practical significance. Thus, if depth to a hard, coherent substrate is chosen as a property differentiating classes, its potential range would be divided up into maybe three or four classes according to practical cut-offs based on say, desired rooting depths for different plant species. In this case, numerical methods of classification would not necessarily be useful as the taxonomic chops are based on practical rather than statistical considerations or natural clustering. The advantage of this type of classification is that it has a direct application. The disadvantage is that it is an artificial construct that often does not coincide with the way soils are distributed in the field. Problems of representation on maps can arise where a discrete area of soils contains a very narrow range of a particular property which happens to span a class differentiating boundary

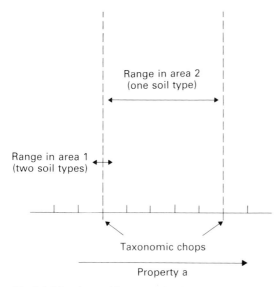

Fig. 2.3 Mapping problems resulting from the use of taxonomic chops.

(Fig. 2.3). Two 'different' soil types could therefore be recognized but not separated within the area and it is difficult to know the best way of representing this situation on a map. If the area is shown as containing two different soil types, this misrepresents the inherent similarity of its diagnostic property. Conversely, if it is shown as a single soil type, then this potentially misrepresents the actual range of the property. To date no entirely satisfactory resolution of this problem has been devised, but it should be stressed that the problem is related to the use of taxonomic chops and the continuously variable nature of soil properties, rather than use of a particular type of classification system.

Hierarchical classifications assist memory and are well suited to the construction of keys to identify individuals. They can also be used at a number of different levels of detail, depending on the amount of information desired. For this reason they are the most commonly used system of soil classification. Hierarchies are, however, particularly susceptible to mapping problems resulting from taxonomic chops (Webster, 1968 and 1977) as two soils separated at the highest categoric level can resemble each other in every respect except the characteristic used for differentiation. Coordinate classification systems are less susceptible to such problems but, unless they are based on a very limited number of properties,

become unwieldy when defined by the possible combinations available. For example, if ten soil properties are used as differentiating criteria, each of which is divided into four classes by three taxonomic chops, a combined total of 4^{10} (1 048 576) classes would be created within a coordinate system.

THE BASIS OF AN URBAN SOIL
CLASSIFICATION

The principle of 'technology transfer', whereby results from detailed local research can be applied to other, often widely-separated areas, provided they are subject to a similar range of environmental factors, is widely used in the fields of agriculture and the natural environment. If this principle is to be usefully applied to urban ecology and land management, then it is clear that one of the main requirements is a practical classification of urban soils based on practically determined taxonomic chops. The choice of differentiating properties within such a classification should be based on factors likely to affect the range of management problems and potential uses of urban land. In addition, because such factors are so diverse, any classification of soils in urban areas should be based on as many relevant properties as is practical. Because of this, a coordinate system will become unwieldy unless extensive computing facilities are available. Even if this is the case, such a system is likely to be 'user unfriendly' in that it will be difficult to visualize and memorize the many different multi-coordinate classes. A hierarchical system would thus be preferable, especially as it has the added attraction of flexibility, enabling information to be extracted at different levels depending on the type and amount required. Nevertheless, the limitations of hierarchical systems must always be borne in mind. Finally, because any classification needs to be applied to the large-scale mapping of urban areas, it needs to be based on properties that are readily evaluated in the field, or that can be inferred from field examination by comparison with 'bench-mark' soils sampled for laboratory analysis (Soil Survey Staff, 1975; Avery, 1980).

Having established some of the basic principles necessary for a useful classification of urban soils, we can now pass on to a more detailed consideration of the types of soil properties that are likely to be needed when differentiating classes.

The soil properties likely to be needed for a useful classification of urban soils

When considering the classification and mapping of tropical soils, Protz (1981) used principles similar to those established in the previous section, to give reasons for choosing soil properties which have an important effect on the growth and production of tropical crops (Table 2.1). In a similar way it is possible to give reasons for choosing a range of properties likely to be of use in any practical classification of urban soils (Table 2.2). As well as stating the reasons for each choice, an assessment of whether the property can be identified in the field and the possible bases for establishing meaningful taxonomic chops are also given. The range of properties considered is based on three main factors: first, the likely differences between urban and non-urban soils; second, the range of urban soil management problems to be reviewed in this publication;

and finally, the range of soil characteristics that affects plant growth.

Few direct comparisons of the differences between urban and non-urban soils have been carried out. Blume (1982) in studying soils of the densely populated area of Berlin, noted that the original soil types had been modified by heavy metal and salt contamination, lowering of the water table, compaction, deep cultivation by gardening and waste disposal. Short *et al.* (1986b), following a statistically-based study of soils in The Mall, Washington DC, found elevated levels of heavy metals, particularly of lead, and abnormally high bulk densities averaging 1.61 t m^{-3} in the surface horizon and 1.74 t m^{-3} at 0.3 m depth. Craul & Klein (1980), investigating street-side soils in Syracuse, New York, concluded that most conditions present in the urban situation tended to destroy soil structure and increase bulk density. During a study of man-induced soil changes in areas adjacent to metallurgical plants in the

Table 2.1 List of ten soil properties important for tropical crops and reasons for their choice (from Protz, 1981)

Soil properties	Reason chosen
1 Slope	Mappable, important from mechanization and conservation viewpoints
2 Effective soil depth	Mappable, important for root penetration, nutrient and water-holding capacity
3 Soil texture – structure	Mappable, important for water release, workability, aeration and root penetration
4 Drainage	Mappable, important for mechanization, aeration, water supply
5 Water release	Mappable from texture and structure, important for crop water uptake
6 Salinity	Can be approximated for soil series, important for plant growth
7 pH	Can be approximated for soil series, important in fertility, minor element deficiencies, toxicity effects of certain elements, plant disease resistance and liming requirements
8 Depth to acid sulphate	Mappable, important due to effects on pH, salt concentration, and root penetration
9 Thickness of peat	Mappable, important for mechanization, tree-crop stands, and shrinkage
10 Workability	Mappable, important for cultivation of annuals and certain tree crops

Table 2.2 Some potential diagnostic properties for a classification of urban soils

Diagnostic property	Reason chosen	Identifiable in the field	Possible basis for taxonomic chops
1 Depth of disturbed or replaced material	Distinguishes between 'natural' and disturbed soils	Can be directly measured	The depth of disturbed material necessary to alter significantly natural soil properties
2 Depth of topsoil	Important for land management, estimation of soil biomass and available water content	Can be directly measured	Critical organic matter levels (see property 4) and important differences in plant productivity, Keleberda & Drugov (1983)
3 Depth of 'weathered' material	Important for estimating the potential for change of soil properties with time	Identified by colour change and structural development	Critical ratios of depth of weathered material to depth of 'exploitable' material
4 Organic matter content	Important for maintenance of structure, nutrient balance and available water content	Can be crudely estimated from colour comparisons and benchmark data	Critical levels for maintenance of structure, nutrient retention and available water
5 $CaCO_3$ content	Important for soil/plant relationships and maintenance of structure	Can be crudely estimated using dilute HCl	Presence or absence and critical levels for plant tolerance
6 Free iron content	Important for maintenance of structure	Can be very crudely estimated by colour	Critical levels for weathering and structural relationships
7 pH	Important for soil nutrient balances; plant growth and liming requirement	Can be directly measured	Relationship between pH and base saturation; nutrient deficiency and species tolerance
8 Cation exchange capacity/Base saturation/ Exchangeable bases	Important for assessing fertility status, nutrient potential and acid-buffering capacity	Can be crudely estimated from texture, organic matter content and pH	Potential for the release or depletion of bases within a fixed period
9 Concentration of toxic minerals, chemicals or gases	Important for land reclamation, ecological planning and human health	Not usually identifiable using field characteristics, but can sometimes be crudely estimated from site history	Threshold values for contamination, ICRCL (1983); Hawley (1985); Bell (1985); Moen *et al.* (1986)
10 Soluble salt concentration	Important for land reclamation, ecological planning and plant growth	Can be directly measured using a conductivity probe	Critical plant-tolerance levels
11 Saturated hydraulic conductivity	Important for plant growth, aeration, land reclamation, soil handling and water supply	Can be crudely estimated from density, texture, macroporosity, structure and gley morphology	Relationships between hydraulic conductivity and duration of soil waterlogging or plant moisture supply
12 Air capacity	Important for plant and animal growth	Can be estimated from macroporosity, structure, density and gleying	Critical levels for plant growth

Table 2.2 (*continued*)

Diagnostic property	Reason chosen	Identifiable in the field	Possible basis for taxonomic chops
13 Dry bulk density	Important for rooting, potential permeability, aeration and available water	Can be estimated from texture, structure and gleying	Critical levels for root growth (Thompson *et al.*, 1987), permeability and faunal activity (Armstrong & Bragg, 1984)
14 Profile available water	Important for plant growth	Can be estimated from texture, bulk density and stone content	Critical available water requirements of different plants (Craul, 1985)
15 Handling properties (plasticity index, shrinkage potential)	Important for land management, reclamation and landscaping	Can be crudely estimated from texture, density, $CaCO_3$ content and free iron content	Critical plasticity indices for soil-handling and shrinkage potentials
16 Duration of soil waterlogging	Important for soil aeration, plant growth and soil handling	Can be crudely estimated from texture, structure, density, gley morphology and climate (Field Capacity Day period)	Relationship between average growing season and duration of waterlogging, or, for soil handling, on absolute values
17 Depth to a hard, coherent substrate	Important for the estimation of potential rooting volume, soil permeability and land management	Can be directly measured	Critical rooting requirements of different plants
18 Nature of the substrate	Important for the estimation of potential toxicity, potential nutrient available and permeability	Can be directly assessed	Critical values of substrate permeability, potential nutrient supply and potential toxicity
19 Vegetative cover	Important for the ability to protect soil from erosion and surface capping	Can be directly assessed	Critical values for reducing rainfall impact

Soviet Union, Fedorishchak (1978) noted sulphate accumulation, a decrease in organic matter content, the inclusion of large amounts of anthropogenic material, the accumulation of heavy metals and an unusually diverse soil pattern because of physical disturbance.

Other comparisons of natural and disturbed soils outside urban areas show fewer differences, at least in some physical properties. Blume *et al.* (1983), comparing related groups of soils on and near a waste disposal site, found no significant differences in erosion/sedimentation processes, gas constituents or the composition of infiltration water. In a study of soil properties on natural and reclaimed hillslopes, Toy & Shay (1987) reported no significant differ-

ences in bulk density, gravel content, pH and organic matter in the upper 10 cm of soil material but, on average, pH and organic matter below 10 cm depth were significantly lower. King (1988) compared the soil physical properties of a restored open-cast coal site in Northumberland with the surrounding undisturbed soil and found that bulk density was significantly higher in the upper subsoil layer (starting at 30 cm depth) of the restored subsoil. In addition there were significantly fewer biotic channels and significantly weaker soil structure in the restored soils below 30 cm depth.

Climatic differences between urban and non-urban environments are better documented (Craul, 1985; Barrett, 1987). They have been summarized

Table 2.3 Climatic effects of urbanization (after Lansberg, 1981)

Element	Comparison with rural areas
Contaminants	
Condensation nuclei	10 times more
Particulates	10 times more
Gaseous admixtures	5–25 times more
Radiation	
Total horizontal surface	0–20% less
Ultraviolet – winter	30% less
Ultraviolet – summer	5% less
Sunshine duration	5–15% less
Cloudiness	
Clouds	5–10% more
Fog – winter	100% more
Fog – summer	30% more
Precipitation	
Amounts	5–15% more
Days with less than 5 mm	10% more
Snowfall – inner city	5–10% less
Snowfall – lee of city	10% more
Thunderstorms	10–15% more
Temperature	
Annual mean	0.5–3.0°C more
Winter minima (mean)	1–2°C more
Summer maxima	1–3°C more
Heating degree days	10% less
Relative humidity	
Annual mean	6% less
Winter	2% less
Summer	8% less
Wind speed	
Annual mean	20–30% less
Extreme gusts	10–20% less
Calms	5–10% more

by Lansberg (1981) and are shown in Table 2.3. The differences must produce associated changes in the temperature and moisture regimes of urban soils but few, if any, site studies have been undertaken. Finally, both Craul (1985) and Barrett (1987) have reviewed the differences that can occur between urban and non-urban soils and environments. These can be summarized as follows:

1 Higher temperature and rainfall, with lower wind speed, relative humidity and radiation. This gives differences in the potential soil moisture deficit and field capacity periods which will, in turn, affect soil temperature and moisture regimes.

2 The extensive presence of man-made materials, either as the physical component of soil, or as extra chemical inputs such as salts, heavy metals, etc.

3 Different physical properties, usually more compact topsoils and/or subsoils and surface crusts on bare soil. This tends to restrict aeration and the downward percolation of surface water. Again soil moisture regimes will be affected.

4 Interrupted nutrient cycling and modified soil organism activity.

5 Greater and less predictable vertical and spatial variation of properties because of human activity. This occurs either directly from local disturbance, removal or replacement of soil material, or indirectly as a result of local physical and chemical alteration of the upper few centimetres of soil.

The greater, and less predictable, variability of urban soils has two important consequences. Firstly, it means that in any useful urban soil classification, a clear distinction between disturbed and undisturbed soils is necessary and, if the system is hierarchical, should be made at a high level. This is because, in undisturbed soils, most of the diagnostic properties used to separate classes will be 'morphogenetic' (acquired as a result of soil-forming processes acting on different soil-parent materials over a period of time). In disturbed soils, however, such properties are likely to be non-existent, very weakly expressed or, if the disturbed material is derived from pre-existing soils, misleading and they will thus be of little use in separating classes. Secondly, it means that in most cases, detailed soil sampling and small-scale mapping will be necessary to establish useful soil patterns in urban areas.

Other papers in this book are devoted to the specific problems of soil contamination, the presence of waste materials, soil storage and handling, urban soil physical properties, urban soil biological properties, nutrient provision and cycling and soil/vegetation relationships. These subjects are therefore not dealt with in detail here, but have been used as a basis for selecting many of the potential diagnostic properties listed in Table 2.2.

The range of urban soil characteristics that affect plant growth, as summarized by Craul (1985), are structure and density, aeration and drainage, available water, permeability, potential rooting volume and configuration, soil reaction and fertility status, and surface protection cover. Any direct assessment of these characteristics should be considered as potentially useful for a classification of urban soils and has therefore been included in Table 2.2

Of the 19 properties listed in Table 2.2, only the level of toxic minerals cannot be directly assessed or measured in the field, or estimated from other properties. However, estimation of free iron content, cation exchange capacity/base saturation/exchangeable bases, saturated hydraulic conductivity, duration of waterlogging, air capacity, handling properties, bulk density and profile available water, all require experience and the use of simple models based on extensive laboratory data. Further, with the exception of profile available water, only broad classes of each property, rather than specific values, can be estimated. For instance, it is possible to assign soil layers to one of several classes of hydraulic conductivity using field assessments of density, structure, macroporosity and gley morphology (McKeague *et al.*, 1982; Wang *et al.*, 1985; McKeague & Topp, 1986; Watt *et al.*, 1986). Similarly, classes of cation exchange capacity, base saturation and exchangeable bases can be predicted using texture and pH (Wang & Coote, 1981). In the case of profile available water, specific values of total available water (held between suctions of 5 and 1500 kPa) or of 'easily available water' (held between suctions of 5 and 200 kPa) can be estimated using field assessments of texture, bulk density and stone content (Hollis, 1987).

All the properties given in Table 2.2, with the possible exception of the level of toxic minerals, thus satisfy the basic principles necessary for a satisfactory practical classification of urban soils. They can therefore be used to assess the effectiveness of existing soil classification systems for this purpose.

Existing classifications and their usefulness in the urban context

Two main factors are used to assess the usefulness of existing classification systems in the urban context. Firstly, there should be a clear separation between relatively undisturbed soils in which morphogenetic properties can be used to make predictions about soil behaviour and those soils which, as a result of man's activities, either have no morphogenetic properties, or have morphogenetic properties derived from pre-existing soils. Secondly, the range of diag-

nostic properties used to differentiate classes within both types of soils described above, should include many of those listed in Table 2.2.

In this kind of review, it is neither possible nor relevant to try and cover the multitude of classification schemes that have been developed for use in specific countries or proposed over the years in journals and other texts. Only those schemes which make some sort of distinction between natural or semi-natural soils and those produced or substantially modified by man are considered relevant to a classification of soils of urban areas and are discussed below.

CLASSIFICATION OF DERELICT LAND IN GREAT BRITAIN

Although not specifically applied to soil, these classifications deal mainly with the urban environment and usually contain categories based on man-made soil substrates or sources of soil contaminants. For this reason they are included here. Their development is fully reviewed by Bridges (1987), upon whose work the following summary is based.

The first comprehensive classification of derelict land in Great Britain was proposed by Beaver (1946) (Table 2.4). It is a multi-ordinate system based on the separation of subclasses within four main properties: surface relief (nine classes); drainage (four classes); vegetation cover (six classes) and composition of the surface (substrate type) (nine classes). Subsequently, Beaver proposed the addition of two further properties: appearance of the surrounding area (12 classes) and ease of restoration (four classes). Oxenham (1966) proposed a much simpler coordinate classification based on land-use or drainage subdivisions within two basic types of topographical situation, namely mounds and spoil heaps (ten land-use classes) and pits and excavations (four classes). A more detailed classification was developed by Collins & Bush (1969) for use with aerial photographic surveys. Again it is a multi-ordinate system based on four categories: general topography (four classes); pictorial (land-use) description (nine classes); associated activity (28 classes – see Table 2.5) and type of filling materials (28 classes – see Table 2.5). Wallwork (1974) and Downing (1977) produced similar multi-ordinate classifications based on a similar range of properties. Downing's check list is particularly comprehensive in the categories of

Table 2.4 Beaver's (1946) four-digit classification of derelict land

Relief of surface (first digit)
1 More or less level
2 Gently sloping (i.e. gradient not more than 1 in 20)
3 Steeply sloping (i.e. gradient over 1 in 20)
4 Level, but pot-holed by subsidence
5 Irregular mounds and hollows (amplitude under 10 ft)
6 As 5, but amplitude over 10 ft
7 Large spoil banks, projecting from surface – conical
8 As 7, but other shapes
9 Marl holes, quarries

Drainage (second digit)
1 Permanently waterlogged
2 Liable to flood
3 With waterlogged hollows
4 Generally free of surface water

Vegetation (third digit)
1 Bare, little or no vegetation
2 Mainly weeds
3 Weeds with sufficient grass to provide a scanty grazing
4 Grass, weeds, and brambles or other small bushes
5 Mainly trees
6 Trees

Composition of the surface (fourth digit)
1 Shale
2 Burnt shale
3 Shale mixed with stones and other debris
4 Stones or quarry waste
5 Blast furnace or other slag
6 Rubble, bricks, concrete, etc.
7 Chemical waste
8 Ashes and cinders
9 Town or other domestic refuse

Table 2.5 Classifications of associated activity and type of filling materials (Collins & Bush, 1969)

Associated activity – general	Associated activity – specific	
1 Mineral working	i	Coal
	ii	Brick clay
	iii	Lead
	iv	Ironstone
	v	Limestone
	vi	Chalk
	vii	Sand and gravel
	viii	China clay
	ix	Tin
	x	Slate
	xi	Others
2 Refuse	i	Household waste
	ii	Scrap
	iii	Cars
	iv	Others
3 Industrial workings	i	Brickworks
	ii	Chemical works
	iii	Gas works
	iv	Iron and steel works
	v	Power stations
	vi	Sewage works
	vii	Others
4 Transportation	i	Airfields
	ii	Canals
	iii	Railways
	iv	Roads
	v	Others
5 Others	i	Others

Filling materials – general: Refers to the nature of the material used to fill an excavation. The categories are described by the numbers 1–5 listed in 'Associated activity – general' above.

Filling materials – specific: The categories are described by the Roman numerals listed in 'Associated activity – specific' above.

'cause of despoliation' (34 classes) and vegetational cover (seven classes) (see Table 2.6). Finally, Haines (1981) gives a very simple classification of contaminated land based on eight classes assessing the extent and potential of contamination, whilst Bridges (1987), in Chapters 3 and 4 of his book on surveying derelict land, gives an inventory of the problem components of derelict and contaminated land along with guidelines as to which components can be expected within land formerly occupied by each of ten types of industrial sites defined by the Central Statistical Office (1980).

None of these classifications defines soil types, but some of their components, such as the drainage, vegetation cover or substrate type classifications given by Beaver (1946), Collins & Bush (1969) and Downing (1977), or Bridges' (1987) list of potential problems and contaminants, could be usefully modified and incorporated into a classification of urban

Table 2.6 Classifications of cause of despoliation and vegetational cover from Downing's (1977) check-list

Cause of despoliation (general)	Cause of despoliation (particular)	Vegetational cover
1 Mineral working	a Chalk b China clay c Clay and shale d Coal e Gypsum/anhydrite f Igneous rock g Ironstone h Limestone j Sand and gravel k Sandstone l Silica and moulding sands m Slate n Vein mineral o Other minerals	1 None 2 Sparse 3 Ephemerals 4 Herbs only 5 Herbs and shrubs 6 Herbs and trees 7 Shrubs and trees only
2 Tipping	a Public waste b Commercial waste c Others	
3 Industry	a Brickworks b Chemical works c Gasworks d Iron and steel e Power station f Sewage works g Others	
4 Transport	a Airfields (private) b Canals c Railways (British Rail) d Roads e Others	
5 Services (military, etc.)	a Airfields (military) b Camps c Defence establishments d Others	
6 Others	a Others (specify)	

soils. Other components, such as the classification of surface relief or source of dereliction, have no place in a soil classification, but could be used to complement or add to the classification of urban habitats proposed by Barrett (1987). This, together with a properly developed soil classification for urban areas, would provide two of the three bases necessary for a comprehensive reference framework for urban environmental research.

THE SOIL CLASSIFICATION FOR ENGLAND AND WALES

This is a hierarchical system with four categorical levels: major soil group; soil group; soil subgroup and soil series. It is fully described by Avery (1980) and Clayden & Hollis (1984). At the highest level, a major group of man-made soils is recognized and defined by the two following diagnostic properties.

First, at least half of the upper 80 cm consists of mineral material, or there is less than 30 cm of organic material resting directly on extremely stony material. Second, either there is a dark 'man-made' A-horizon at least 40 cm thick that contains at least 0.6% organic carbon throughout, with the organic matter intimately mixed with the mineral fraction and which usually contains artefacts such as brick or pottery fragments; or there is a disturbed subsurface layer that extends below 40 cm depth that consists wholly or partly of materials derived from pedogenic horizons of pre-existing soils.

At the group level man-made soils are subdivided into *man-made humus soils* characterized by a dark man-made A-horizon at least 40 cm thick, and *disturbed soils* characterized by the presence of a disturbed subsurface layer that extends below 40 cm depth. Man-made humus soils are further separated into *sandy* and *earthy* subgroups according to the predominant texture within the upper 80 cm of their profiles. Disturbed soils, however, are not subdivided below group level, although on the 1:250 000 scale soil maps of England and Wales (Soil Survey of England and Wales, 1984), three types are distinguished: disturbed soils on restored opencast coal sites; on restored ironstone workings; and on restored coprolite (phosphatic nodules) workings. Further separations of disturbed soils on land restored after the extraction of sand, gravel or brick earth have also been suggested (Avery, pers. comm.).

The England and Wales system makes a useful separation of many man-influenced soils, but it should be noted that soils with surface or subsurface horizons which appear to have originated naturally in or on artificially emplaced material not previously affected by soil-forming processes would not be classed with man-made soils. Instead they would be placed within other major soil groups depending on their morphogenetic diagnostic properties. Because of this, the name man-made soils is unfortunate as the corresponding soils in other major soil groups could undoubtedly be described as being 'man-made'.

Soils that are formed in artificially emplaced material, but do not qualify as man-made soils, are subdivided using some of the properties listed in Table 2.2, for example: depth and organic matter content of the topsoil; depth of weathered material;

$CaCO_3$ content; permeability and duration of waterlogging. Along with depth of disturbed material, depth and organic matter content of the topsoil are also used to separate classes within man-made soils, but other properties are not. Important diagnostic criteria are thus not applied consistently in the England and Wales system which limits its use in the urban context.

USA SOIL TAXONOMY

This is by far the most detailed and comprehensive soil classification system in existence today. It was developed by the Soil Conservation Service of the United States Department of Agriculture (Soil Survey Staff, 1975) and its stated purpose is to provide 'a basic system for making and interpreting soil surveys'.

The system is a hierarchical one with class distinctions based on precisely defined, diagnostic properties often combined as 'diagnostic horizons'. Classes of soil moisture regimes and soil temperature regimes are incorporated at two different levels. In contrast to many other classifications which use everyday words, or well-established, but often badly or ambiguously defined soil terminology to name classes, it uses a completely new nomenclature derived from Greek and Latin, the comparison with biological taxonomy being obvious. There are six categorical levels: Order; Suborder; Great group; Subgroup; Family and Soil series. A formative element from names in each of the three highest categories is carried down to form subgroup names and, at the family level, textural, mineralogical and soil temperature regime names are added (Table 2.7). At the lowest level, soil series are simply named from a geographic location.

Table 2.7 Nomenclature related to a soil class in *Soil Taxonomy* (Soil Survey Staff, 1975)

Order	Spodosols
Suborder	Orthods
Great Group	Fragiorthods
Subgroup	Typic Fragiorthods
Family	Typic Fragiorthod, coarse-loamy (particle-size), mixed (mineralogy), frigid (temperature regime)

Table 2.8 Soil orders and their diagnostic properties in *Soil Taxonomy* (Soil Survey Staff, 1975)

Soil order	Mnemonic	Diagnostic properties	Type of soils meant to be included
Entisols (-ent)	Recent	Diagnostic subsoil horizons are absent	Weakly developed or young soils
Vertisols (-ert)	Invert	With vertic properties (clayey texture with deep, wide cracks at some time in most years) and either gilgai features or slickensided or wedge-shaped subsoil structures	Self-mulching or mixing, cracking-clay soils
Inceptisols (-ept)	Inception	Lacking any diagnostic subsoil horizon other than a cambic (slightly weathered) one	Moderately developed, slightly weathered (brown) soils of humid regions
Aridisols (-id)	Arid	With an Aridic (dry) moisture regime and an argillic, natric, salic, petrocalcic, calcic, gypsic, petrogypsic or cambic subsoil horizon or a duripan	Soils of deserts and semi-deserts
Mollisols (-oll)	Mollify	With a Mollic epipedon (a well-structured, dark-coloured, base-rich surface layer more than about 15 cm thick)	Base-rich soils of grassland and steppe regions
Spodosols (-od)	Podzols	With a Spodic (podzolic) subsoil horizon enriched in iron, aluminium and/or humus	Podzolized soils
Alfisols (-alf)	Pedalfer	With an Argillic (clay-enriched) subsoil horizon that is at least 35% base-saturated	Soils with significant amounts of clay-illuviation
Ultisols (-ult)	Ultimate	With an Argillic subsoil horizon that is less than 35% base-saturated	Deeply weathered, clay-illuviated soils usually of the humid sub-tropics
Oxisols (-ox)	Oxide	With an Oxic subsoil horizon (with a low cation-exchange capacity and less than 10% 'weatherable' minerals) or Plinthite (iron-rich, humus-poor mixture, mainly of clay and quartz) within 40 cm depth	Deeply weathered, usually reddish (lateritic) soils of the tropics
Histosols (-ist)	Histology	With significant amounts of organic material in the upper parts of the profile	Peaty or organic soils

There are ten soil orders, differentiated mainly by the presence or absence of diagnostic horizons (Table 2.8). Soils highly influenced by man are principally separated at suborder levels either within Entisols or Inceptisols:

1 *Arents* are defined as having, between 25 and 100 cm depth, fragments of diagnostic horizons in no discernible order. They are not permanently saturated with water and do not show characteristics associated with wetness.

2 *Plaggepts* have a man-made surface layer more than 50 cm thick that has been produced by long, continued manuring.

In addition, some *Anthropogenic* subgroups are recognized where a relatively thick, dark-coloured topsoil containing at least 0.6% organic carbon and more than 250 ppm citric-soluble P_2O_5 has developed because of long, continued use by man.

As Clayden (1982) has pointed out, *Soil Taxonomy* is a monumental publication of over 700 pages and it is only possible to classify a soil with confidence by careful reference to the text or to specially prepared keys (Soil Survey Staff, 1987). Criticisms of the system have been voiced by Raeside (1961), Gerasimov (1964), Mulcahy & Humphries (1967), Webster (1968), and Duchaufour (1982), among others, and its advantages and disadvantages have been reviewed by Ragg & Clayden (1973) and Clayden (1982). There is no point repeating such arguments here, but it is worth emphasizing that *Soil Taxonomy* has had a profound influence on many other national schemes, as well as that developed by FAO/ UNESCO for the *Soil Map of the World*. Furthermore, although originally intended as a national system for the USA, it is increasingly being used elsewhere, particularly in the 'developing countries'. This is due mainly to the work of the Soil Management Support Services, a programme of international technical assistance run by the USDA Soil Conservation Service and funded by the US Agency for International Development. To encourage its use as an international reference system and as a basis for agrotechnology transfer, a number of international committees have been formed to develop or modify diagnostic criteria and class differentiae for soil types that do not occur, or are of limited extent, in the USA.

In the urban context, *Soil Taxonomy* suffers from the same kind of limitations that apply to the system for England and Wales. Because of these, soil scientists working in urban areas or on man-influenced soils have proposed a number of amendments, none of which has as yet been officially endorsed.

Sencindiver (1977), discussing the classification and genesis of minesoils, proposed the creation of an extra suborder of *Spolents* within Entisols. Differentiation would be based on up to nine diagnostic properties including the presence of artefacts, irregular distribution of organic matter with depth and the random orientation of coarse fragments. In a paper discussing the genesis and classification of highly man-influenced soils, Fanning *et al.* (1978) state that, in their estimation, 'attempts to place (squeeze) soils representing genetic situations highly influenced by man, into present *Soil Taxonomy* classes have resulted in poor groupings'. To improve this they proposed the creation of four new subgroups to differentiate classes mainly within Orthents and Ochrepts.

The four subgroups are based on genetic and morphological situations highly influenced by man and are:

1 *Scalpic*: differentiated where natural topographic contours are broken and relatively unweathered material approaches the surface.

2 *Garbic*: distinguished where there is organic garbage and/or high methane in the soil atmosphere within 1–2 m of the soil surface.

3 *Urbic*: where inorganic industrial artefacts are present within the control section.

4 *Spolic*: where locally derived earthy spoil without industrial artefacts forms the substrate material.

They also recognize the possibility of creating an extra soil order of *Potisols* (potential soils) which, in their present state are not capable of supporting plants, but have the potential to do so by correction of the problem that prevents plant growth. Raw sulphidic material such as pyritic mine spoil is quoted as an example that could be classed as a *Sulfoudot* (Sulphuric, Udic, Potisol), if it had a udic (moist) moisture regime.

Following soil surveys on surface-mined land in Perry County, Ilinois, Indorante & Jansen (1984) suggested that by creating soil series based on the method of soil reconstruction, they could predict the distribution of soil types which have differences in texture, stoniness, reaction class and density/permeability. For this type of land at least, providing new series-differentiating criteria can be accepted, there is no need to create new classes at higher levels in the system. In a study of 24 non-cultivated, non-topsoiled minesoils in Pennsylvania, Ciolkosz *et al.* (1985) concluded that most would qualify as Entisols but about one-fifth were Inceptisols in which weak, cambic (weathered) subsoils had developed. They supported the creation of a Spolent suborder suggesting that it is important to recognize such soils as being created by man and point to the differentiation of Arent and Plaggept suborders as a precedent.

Short *et al.* (1986a) discussed the problems of using *Soil Taxonomy* to classify and map urban soils in the Mall, Washington. They pointed out that in some cases, even in disturbed materials, weak cambic subsoil horizons have developed but particular problems occur where there is irregular distribution of organic matter with depth. This commonly results in soils being classed in Fluv-subgroups even though they are not developed in recent alluvial material. They concluded that present diagnostic criteria in *Soil Taxonomy* do not adequately differentiate man-influenced soils so as to provide accurate and useful information about them. Use of criteria that will ensure this should be promoted and they supported the recognition of Urbic and Spolic subgroups as proposed by Fanning *et al.* (1978).

Finally, Kosse (1986) proposed the creation of a completely new order of Anthropogenic soils, to isolate a unique range of man-influenced soils and to focus attention on the unity of pedogenic processes involved in their creation. He gives tentative definitions of some important suborders including Plaggans, Hortans, Aquans and Irrigans and suggests that the number of diagnostic horizons be increased to enable the separation of a wider range of Anthropogenic soil classes.

In summary, all the proposals for amendments to *Soil Taxonomy* suggest the creation of additional diagnostic properties to separate man-influenced or man-made soils, but there is disagreement as to the level at which they should be applied. Suggestions include the creation of new orders or suborders or subgroups or soil series. As has been discussed previously, it is best in the urban context to separate man-modified soils at higher levels in hierarchical systems. The proposal for additional orders or suborders are thus of most relevance, although the suggested diagnostic properties include few of those listed in Table 2.2. To be of more relevance to the classification and mapping of soils in urban areas, these proposals would need to be developed using additional diagnostic properties from Table 2.2 to differentiate classes at lower categoric levels.

FAO/UNESCO

Classes of man-influenced soils are only recognized in the *Revised Legend to the Soil Map of the World* (Food and Agriculture Organization, 1988), a development and revision of the original FAO/UNESCO legend produced to accompany the *Soil Map of the World* (Food and Agriculture Organization, 1974). The classification system used in this legend is hierarchical but only has two categorical levels. Classes are defined by diagnostic properties and horizons, many of which mirror those of *Soil Taxonomy*. At the highest level, a class of *Anthrosols* is defined as 'soils in which human activities have resulted in profound modifications of the original soil characteristics, through removal or disturbance of surface horizons, cuts and fills, secular additions of organic materials, long-continued irrigation, etc.'. This is subdivided into four subclasses:

1 *Aric Anthrosols* show only the remnants of diagnostic horizons due to deep cultivation.

2 *Cumulic Anthrosols* show an accumulation of fine sediments more than 50 cm thick, resulting from long-continued irrigation or man-made raising of the soil surface.

3 *Fimic Anthrosols* have a man-made surface layer at least 50 cm thick produced by long-continued manuring with earthy admixtures.

4 *Urbic Anthrosols* have an accumulation of wastes from mines, town refuse, urban fills, etc. to a depth of at least 50 cm.

The FAO system is not meant to be used for detailed soil mapping and thus only makes some broad distinctions between man-influenced soils and more naturally developed ones based on depth of disturbed or replaced material, depth of topsoil and depth of weathered material. As with *Soil Taxonomy* therefore, it is of little practical use for classifying soils in an urban context unless it is developed to lower categorical levels using some or all of the diagnostic properties given in Table 2.2.

THE NETHERLANDS

The current Dutch system (De Bakker & Schelling, 1966) is hierarchical with four levels: Order, Suborder, Group and Subgroup. It is, however, unlike most other soil classifications discussed here in that it was specifically designed for a very limited but special range of soils and landscapes. It is also of interest in that, understandably in this land of man-made soils, it was one of the first to place emphasis on soil properties and class distinctions based on man's influence.

Soils with man-made A-horizons more than 30 cm thick are recognized and separated from otherwise similar soils in the Podzol and Earth orders. A basic distinction is also made between soils with moderately thick (30–50 cm) and thick (75 cm) man-made A-horizons. Reworked soils, which have either been physically disturbed to a depth of more than 40 cm, or consist of at least 20 cm of disturbed material extending to below 40 cm, are defined to distinguish normally cultivated soils from deeply cultivated ones. However, the properties are not used to separate classes in the upper four categories, but instead used to delineate reworked soils on soil maps, where they carry an additional symbol after the map unit code. This is a 'phase' type of distinction.

Although the Dutch system separates many of the basic disturbed or man-made classes of man-influenced soils used in other systems, it does not do so consistently. In addition, 'raw' man-made soil materials present in urban environments are not differentiated at all. For both these reasons and because it was designed for such a limited range of soils and uses, the system is of limited value in urban situations.

WEST GERMANY

The latest *Soil Classification of the Federal Republic of Germany* (DBG, 1985), edited by the German Society of Soil Science working group on soil systematics, is a hierarchical system with seven categorical levels: Order, Suborder, Type, Subtype, Variety, Subvariety and Soil Series. At the highest level there are five orders: Terrestrial soils, Semi-terrestrial soils, Semi-subaqueous and Subaqueous soils; Bog (peat) soils and Periglacial soils. All soils 'whose profile composition has been altered so strongly by direct human action that the original soil horizon sequence has been mostly destroyed' are separated as *Anthropogenic* suborders, but so far these have only been recognized within the Terrestrial and Semi-terrestrial orders. Five types of Anthropogenic Terrestrial soils are described:

1 *Plaggen soils* which have a plaggen layer (formed as a result of prolonged manuring with sods cut from heath or grassland and then composted or used for bedding for livestock) more than 40 cm thick. Subtypes are based either on the colour of the plaggen layer or on the permeability and gley morphology of subsoil layers.

2 *Hortisols* which have an A horizon (topsoil) more than 40 cm thick, formed as a result of intensive tillage and addition and mixing of organic matter over many years, overlying naturally formed soil horizons. The implication is that the topsoil has been 'built up' over the subsoil horizons rather than created by their destruction. Subtypes are based on the thickness of A horizons and their organic matter content, in combination with the original parent soil type as recognized from the subsoil horizons.

3 *Rigosols* which consist of a mixture of pre-existing soil horizons more than 30 cm thick formed as a result of recurrent trench ploughing (Rigolen) which has destroyed the coherency of subsoil horizons. Subtypes are based on the original soil type as recognized from the remaining fragments of original soil horizons.

4 *Deeply disturbed soils* (Treposols) which also consist of a mixture of pre-existing soil horizons more than 30 cm thick, but in this case formed as a result of deep ploughing. They often truncate and overlie otherwise undisturbed subsoil horizons. Again, subtypes are based on the original soil type.

5 *Made-ground soils* which consist of more than 80 cm of raw soil materials largely unaffected by soil-forming processes and formed as a result of man's activities. Subtypes are not described 'because the various soils need individual description' but distinctions based on the type of raw soil material present are possible within the system.

Within Anthropogenic Semi-terrestrial soils, Plaggen, Hortisol and Rigosol Types are recognized and differentiated using similar properties to those used within Anthropogenic Terrestrial soils.

The latest West German system gives a logical and systematic separation of man-influenced soils in its upper categorical levels, based mainly on depth of disturbed or replaced material, depth of topsoil, depth of weathered material, organic matter content and, very broadly, duration of waterlogging. The morphological distinction between Rigosols and Treposols is, however, unclear and, if based solely on differences in the method of deep disturbance, is probably unnecessary for practical purposes. Furthermore, although Plaggen subtypes of non-Anthropogenic suborders are recognized where Plaggen layers less than 40 cm thick occur, no provision is made for Made-ground soils formed on raw man-made soil materials 80 cm thick or less.

With respect to urban soils therefore, there are some minor inconsistencies at higher levels in the system. Details of differentiae below subtype level are not given, but classes based on $CaCO_3$ content, free-iron content, salt concentration, permeability/density and nature of the substrate are provided for.

EAST GERMANY

The soil classification system developed in the German Democratic Republic (Ehwald *et al.*, 1966; Ehwald, 1968) is a coordinate one, where classes are created by combining the name of a soil type or subtype, with a lithological class called a 'Sippe'. The types and subtypes are defined on the basis of pedological characteristics and are derived from an earlier West German system (Muckenhausen, 1962, 1965), but with fewer classes.

Lithological classes are defined according to the nature and origin of soil-parent materials and also on whether they are homogeneous or stratified. The system separates *Plaggen soils*, *Hortisols* and *Rigosols* as in the West German system described above, but deeply disturbed and made-ground soils are not recognized. To cater for some of these omissions, Legler (1970) made some proposals regarding the classification of recultivated areas used in agriculture. He defined a suborder of *Recultisols* developed on 'tailings' (spoil) now used for agriculture or forestry. Subdivision to subtype level is based on land use (agriculture or forestry), substrate thickness suitable for plant growth, $CaCO_3$ content and humus content, whilst lithological classes are defined using other physical and chemical characteristics of the substrate.

Although limited in scope, Legler's proposals use many of the potential diagnostic properties identified in Table 2.2 and are a useful development of the East German system to cover a particular range of man-influenced soils.

YUGOSLAVIA

In a paper discussing the classification of 'Damaged soils', Antonovic (1986) recognized their widespread occurrence and the need to incorporate them into modern classification systems. With reference to the East German system, or an adaptation of it, he proposed recognition of a class of Damaged soils to be separated from existing Anthropogenic classes such as Rigosols, Hortisols and Plaggen soils, but gave no details as to the differentiating criteria. Damaged soils would be subdivided hierarchically at four levels, Type, Subtype, Variety and Form (Table 2.9). Again, no details of diagnostic properties were given but the basis appeared to be similar in some ways to the 'cause of despoliation' or 'type of filling materials' categories within some of the classifications of derelict land discussed earlier (Collins & Bush, 1969; Downing, 1977).

As it stands, Antonovic's proposal is of limited value for soils in urban areas as it lacks detail and includes some 'land-use' classes conceptually incompatible with a morphologically-based classification system. However, it could provide a useful basis for

development as, like many of the derelict land classifications, it makes comprehensive subdivisions on the nature and origin of man-made substrates or pollutants. It is also one of the few schemes to suggest the separation of industrially polluted, but otherwise undisturbed soils (Aerosols), which are likely to be of great significance in urban areas.

USSR

The ecological–genetic approach to soil science, which relates soil properties to pedogenic soil-forming processes or factors, was pioneered by the Russian workers Dokuchaiev and Sibirtsiev. Modern soil classification in the Soviet Union reflects these classic roots in retaining a strong emphasis on ecological-bioclimatic zones and the use of soil-forming processes and environmental factors to differentiate classes, rather than intrinsic soil properties that can be readily observed or measured. The most recent comprehensive scheme published is described by Rozov & Ivanova (1967). It is rather complex, using three coordinate axes to define about 110 genetic soil types, which are then further differentiated at three hierarchical levels mainly using horizon thicknesses, the chemical content of groundwater and soil-parent material characteristics. The three coordinate axes used to define genetic soil types are:

1 Nine ecological–genetic or bioclimatic classes based on climatic indices.
2 Four genetic orders based on moisture regime.
3 Five biophysicochemical orders based mainly on characteristics of organic matter decomposition, soluble salts, degree of base saturation and cation composition.

No differentiation of man-influenced soils is made within the system but some recent work has recognized this omission and suggested ways it can be rectified. Fedorishchak (1978), studying anthropogenic soil changes around metallurgical plants, recognizes six types of soil: natural or undisturbed; shortened, including eroded (this is presumably partly equivalent to the 'scalpic' subgroup suggested by Fanning *et al.*, 1978); soil destroyed by digging machines; fill or 'mixed' materials; and exposed rock. The translation of his paper gives no details of how these soil types are recognized, but they are used to assess the suitability of land for trees and shrubs in landscaping schemes.

Table 2.9 A classification of damaged soils (Antonovic, 1986)

Type	Subtype	Variety
1 *Deposol* (derived from Latin *deponere* and French *depot*, store)	1.1 Open mining	1.1.1 Coal mining 1.1.2 Copper mining 1.1.3 Lead and zinc mining 1.1.4 Bauxite mining 1.1.5 Magnesite mining 1.1.6 Asbestos mining 1.1.7 Iron mining 1.1.8 Clay mining
	1.2 'Lent' material	1.2.1 Sand and gravel 1.2.2 Soils 1.2.3 Quarries/stone-pits
	1.3 The storage of red mud	
	1.4 The storage of ashes/ cinders	1.4.1 'Dry' drifted/deposited 1.4.2 'Wet' drifted 1.4.3 Dross/slag
	1.5 The dump	1.5.1 Domestic waste material 1.5.2 Building waste material 1.5.3 Industrial waste material
2 *Flotasol* Flotation	2.1 Industrial water deposits 2.2 Flooding deposits 2.3 Irrigation water deposits	For all subtypes 2.1–2.3 1 Pyrites waste material 2 Lead and zinc waste material 3 Coal waste material 4 Acid mining waste material 5 Basic mining waters 6 Caustic soda waste material 7 Hydrolysed aluminium waste material

3 *Urbasols* (from Latin *urbs*, town)	3.1 Town settlements
	3.2 Village settlements
	3.3 Country settlements/dachas
	3.4 Factories
	3.5 Great stadiums
	3.6 Airports
		Separation according to
		  Parks
		  Promenades
		  Recreation centres
		  Jogging paths
		  Lawns
		  Low protected surfaces

4 Aerosols (from Greek *aer*, air)	4.1 Sulphur gas pollution
	4.2 Lead and zinc pollution
	4.3 Copper pollution
	4.4 Sintermagnesite pollution
	4.5 Cement pollution
	4.6 'Flying' ash pollution
		For 4.1–4.4
		  Small
		  Medium
		  Great
		  Extreme
		For 4.5–4.6
		  Small
		  Medium
		  Great

Forms for varieties 1.1.1–1.1.8 and 1.2.1–1.2.3 are:

A group – recultivable: I class – very recultivable
  II class – recultivable
  III class – slightly recultivable

B group – nonrecultivable: IV class – nonrecultivable
  V class – very nonrecultivable
  VI class – extremely nonrecultivable

C group – toxic is for 1.3

Table 2.10 Classification of Technogenic soils (from Keleberda & Drugov, 1983)

Class	Subclass	Group	Subgroup	Genus	Subgenus	Species	Series	Variety
Technogenically transformed or disturbed soils (no details of differentiating criteria given)	These include: (i) Trenchplowed (ii) Filled (iii) Terraced (iv) Planated (v) Recultivated	These are defined according to ecological bioclimatic zones, e.g. steppe, forest (see Rozov & Ivanova, 1967)	These are subdivisions of zonal groups, e.g. northern forest, dry steppe, etc.	Two subdivisions: (i) Soils with at least two layers, including a humified topsoil; (ii) Soils with only one layer that is not a humified topsoil	Subdivisions are based on: (i) Lithology and geochemistry of the substrate, *or* (ii) Soil moisture regime, *or* (iii) Presence of significant amounts of erosion or deflation (no details given)	Subdivided according to: (i) Thickness and organic matter content of humified layer; (ia) <30 cm thick; (ib) 30–50 cm thick; (ic) >50 cm thick; (id) <2% organic matter; (ie) 2–4.1% organic matter; (if) >4.1% organic matter; *or* (ii) The degree of development of subgenus characteristics (no details given)	Subdivided according to: (i) Parent material type; (ia) loess or loess-like; (ib) non-calcareous mixed; *and* (ii) Particle-size: clayey loamy; sandy loamy, etc.	Subdivisions are based on the level (degree) of cultivation: (i) Slightly cultivated; (ii) Cultivated (no details given)

In a comprehensive paper on the systematics and classification of 'Technogenic' soils, Keleberda & Drugov (1983) recognize the importance of studying soil evolution under anthropogenic and technogenic influences. They emphasize the need for this study in order to resolve the many questions arising from utilization of technogenic landscapes. Large areas of disturbed soils have been created within the Soviet Union over the last 20 years and these cannot as yet be included in Russian soil classifications or agricultural groupings. Drawing on much previous work, mainly in Moldavia and the Ukraine (Krupenikov & Podymov, 1973; Polupan, 1981; Denisik & Roychenko, 1982; Yeterevskaya *et al.*, 1982) they elaborate the principles and basis of differentiae for a comprehensive hierarchical classification of Technogenic soils (Table 2.10), but give only sketchy details of many diagnostic criteria.

As the authors recognize, theirs is only an initial proposal that requires additional research and refinement. The scheme contains many undefined concepts and some inconsistencies, possibly arising from difficulties of translation. As with most Russian soil classifications, it also includes zonal concepts which have very little use in the UK. Nevertheless it is the most comprehensively developed system for the classification of man-influenced soils so far published and contains many of the basic criteria and potential diagnostic properties necessary for a useful classification of urban soils.

Conclusions

The preceding review of current soil classification schemes shows that, whereas some use diagnostic properties attributable to man's influence, this is done mainly in an agricultural context to differentiate soils that have been created or modified by unusual agricultural practices such as deep or trench ploughing, or the regular application of various types of manurial material. Even where other kinds of disturbed or man-made soils are recognized, such as in the USDA *Soil Taxonomy* or the system of England and Wales, they are separated at different hierarchical levels, causing confused nomenclature and inconsistent application of diagnostic criteria. Only in the systems developed for West Germany and FAO/UNESCO are there attempts to separate the whole range of man-influenced or modified soils as a single class at a high categoric level. Even here some minor omissions and inconsistencies occur and, at lower levels, precise differentiating criteria have not been fully developed.

In their present form therefore, none of the nationally or internationally used soil classification schemes fulfil all the requirements for a useful classification of urban soils. Of the various amendments to these schemes that have been proposed to extend their application into areas of highly man-influenced soils, the most comprehensive and potentially most useful are those of Keleberda & Drugov (1983), Antonovic (1986), Fanning *et al.* (1978), and Kosse (1986). None could be used as they stand, however, as they all contain at least some omissions, inconsistencies, undeveloped concepts and inadequately defined criteria.

If a useful practical classification of urban soils is to be developed for the UK, it is suggested that the existing soil classification for England and Wales (Avery, 1980) be adapted to separate a major soil group of Anthropogenic soils using diagnostic properties 1 and 4 from Table 2.2. These could then be subdivided into a number of soil groups, similar to those recognized in the West German or FAO systems, using diagnostic criteria based, among others, on properties 1 to 4 in Table 2.2. Subdivisions at the lower hierarchical levels should use diagnostic criteria of practical importance, including many or all of properties 5–19 listed in Table 2.2.

It is further suggested that the range of classes separated should be similar to those proposed by Keleberda & Drugov (1983) and Fanning *et al.* (1978). Classes based on the physicochemical composition of man-made substrates should recognize the range of industrial waste materials encompassed in the classifications of derelict land proposed by Beaver (1946), Collins & Bush (1969), and Downing (1977) or the classification of Damaged soils suggested by Antonovic (1986). Finally, it will be necessary to decide in what major soil group, if any, soils similar to Antonovic's *Aerosol* class (which have been physically or chemically modified but are otherwise undisturbed) should be separated.

References

Antonovic, G.M. (1986). Classification of damaged soils. *Transactions of the 13th Congress of the International Society of Soil Science*, Hamburg, FGR **3**, 1036–1037.

Armstrong, M.J. & Bragg, N.C. (1984). Soil physical parameters and earthworm populations associated with opencast coal working and land restoration. *Agriculture, Ecosystems and Environment* **11**, 131–143.

Avery, B.W. (1968). General soil classification: hierarchical and coordinate systems. *Transactions of the 6th International Congress of Soil Science, Paris, France* **E**, 279–285.

Avery, B.W. (1980). *Soil Classification for England and Wales (Higher Categories)*. Soil Survey Technical Monograph No. 14, Harpenden.

Barrett, I. (1987). *Research in Urban Ecology*. Mimeograph Report to the Nature Conservancy Council.

Beaver, S.H. (1946). *Report on Derelict Land in the Black Country*. Mimeograph from the Ministry of Town and Country Planning, London.

Bell, F.G. (1985). *Engineering Properties of Soils and Rocks*, pp. 69. Butterworths, London.

Blume, H.P. (1982). Boden des Verdichtungsraumes Berlin (Soils of the high density area of Berlin). *Mitteilungen der Deutschen Bodenkundlichen Gesselschaft* **33**, 269–280.

Blume, H.P., Hofmann, I., Mouimou, D. & Zingk, M. (1983). Bodengesselschaft auf und neben einer Mulldeponie (Related groups of soils on and near a waste disposal site). *Zeitschrift fur Pflanzenernahrung und Bodenkunde* **146**, 62–71.

Bockheim, J.G. (1974). Nature and properties of highly disturbed urban soils, Philadelphia, Pennsylvania. *Paper presented before Div. S-5, Soil Science Society of America, Chicago, Illinois.*

Bridges, E.M. (1982). Techniques of modern soil survey. In Bridges, E.M. & Davidson, D.A. (eds), *Principles and Applications of Soil Geography*, pp. 28–57. Longman, London.

Bridges, E.M. (1987). *Surveying Derelict Land*. Monographs on Soil and Resources Survey No. 13, Oxford University Press, Oxford.

Butler, B.E. (1980). *Soil Classification for Soil Survey*. Monographs on Soil Survey, Oxford University Press, Oxford.

Central Statistical Office (1980). *Standard Industrial Classification*. HMSO, London.

Ciolkosz, E.J., Cronce, R.C., Cunningham, R.L. & Peterson, G.W. (1985). Characteristics, genesis, and classification of Pennsylvania minesoils. *Soil Science* **139**, 232–238.

Clayden, B. (1982). Soil classification. In Bridges, E.M. & Davidson, D.A. (eds), *Principles and Applications of Soil Geography*, pp. 58–96. Longman, London.

Clayden, B. & Hollis, J.M. (1984). *Criteria for Differentiating Soil Series*. Soil Survey Technical Monograph No. 17, Harpenden.

Cline, M.G. (1949). Basic principles of soil classification. *Soil Science* **67**, 81–91.

Collins, W.G. & Bush, P.W. (1969). The definition and classification of derelict land. *Journal of the Town Planning Institute* **55**, 111–115.

Craul, P.J. (1985). A description of urban soils and their desired characteristics. *Journal of Arboriculture* **11**, 330–339.

Craul, P.J. & Klein, C.F. (1980). Characterization of streetside soils in Syracuse, New York. *METRIA* **3**, 88–101.

De Bakker, H. (1970). Purpose of soil classification. *Geoderma* **4**, 195–208.

De Bakker, H. & Schelling, J. (1966). *Systeem van bodemclassificatie voor Nederland, De Logere Niveaus*, 217 pp. Pudoc, Wageningen, The Netherlands.

Denisik, G.J. & Roychenko, G.I. (1982). The rational use of technogenic soils in Podol'ye. *Abstracts of papers read at the 1st Delegates' Congress of the Soil Scientists and Agricultural Chemists of the Ukrainian SSR Soil melioration, erosion control and recultivation.* Kharkov.

Deutsche Bodenkundliche Gesellschaft. (1985). *Soil Classification of the Federal Republic of Germany.* (Mitteilungen der Deutschen Bodenkundlichen Gesellschaft). Working Group on Soil Systematics of the German Society of Soil Science (eds). P. Hugenroth, Gottingen.

Downing, M.F. (1977). Survey information. In Hackett, B. (ed.), *Landscape Reclamation Practice*, pp. 17–36. IPC Science and Technology Press, Guildford.

Duchaufour, Ph. (1982). *Pedology: Pedogenesis and Classification*. Translated by T.R. Paton. George, Allen and Unwin, London.

Ehwald, E. (1968). Some new approaches to soil classification in the German Democratic Republic. *Soviet Soil Science* **10**, 1329–1336.

Ehwald, E., Lieberoth, I. & Schwanecke, W. (1966). *Zur Systematik der Boden der Deutschen Demokratischen Republic, besonders in Hinblick auf die Bodenkartierung.* Sitzungsberichte Deutsche Akademie der Landwirtschafts-wissenschaften, Berlin, 15.

Fanning, D.S., Stein, C.E. & Patterson, J.C. (1978). Theories of genesis and classification of highly man-influenced soils. In *Abstracts for Commission Papers, Vol. 1, 11th Congress of the International Society of Soil Science*, pp. 283. Edmonton, Canada.

Fedorishchak, M.R.P. (1978). Anthropogenic soil changes in the zone of influence of metallurgical plants. *Soviet Soil Science* **11**, 133–137.

Food and Agriculture Organization. (1974). *Soil Map of the World, 1:5 000 000*, Vol. 1, Legend, UNESCO, Paris.

Food and Agriculture Organization. (1988). *Soil Map of the World: Revised Legend.* World Soil Resources Report 60, Final Draft, FAO, Rome.

Gerasimov, I.P. (1964). Discussion of the new American soil classification system. *Soviet Soil Science* **20**, 572–603.

Haines, R. (1981). *Contaminated Sites in the West Midlands County. A Prospective Survey.* JURUE, University of Aston, Birmingham.

Hawley, J.K. (1985). Assessment of health risk from exposure to contaminated soil. *Risk Analysis* **5**, 289–307.

Hollis, J.M. (1987). *The Calculation of Crop-adjusted Soil Available Water Capacity for Wheat and Potatoes.* Soil Survey Research Report No. 87/1, Silsoe.

Indorante, S.J. & Jansen, I.J. (1984). Perceiving and defining soils on disturbed land. *Soil Science Society of America Journal* **48**, 1334–1337.

Interdepartmental Committee on the Redevelopment of Contaminated Land. (1983). *Guidance on the Assessment and Redevelopment of Contaminated Land.* ICRCL 61/84. DOE, London.

Keleberda, T.N. & Drugov, A.N. (1983). Systematics and classification of technogenic soils. *Soviet Soil Science* **15**, 61–67.

King, J.A. (1988). Some physical features of soil after opencast mining. *Soil Use and Management* **4**, 23–30.

Kosse, A. (1986). Anthrosols: Proposals for a new soil order. *Transactions of the 13th Congress of the International Society of Soil Science*, Hamburg, FGR **3**, 1175.

Krupenikov, I.A. & Podymov, B.P. (1973). Classification and systematics of Moldavian soils. In *Genezis, Geografiya, i Klassifikatsiya pochv Moldavii.* Kishinev, Shtintsa.

Lansberg, H.E. (1981). *The Urban Climate.* Academic Press, New York.

Legler, B. (1970). Proposals regarding the classification of recultivated areas used in agriculture. In *Symposium on the Recultivation of Regions disturbed by Industry*, Part 1. Leipzig.

McKeague, J.A. & Topp, G.C. (1986). Pitfalls in interpretation of soil drainage from soil survey information. *Canadian Journal of Soil Science* **66**, 37–44.

McKeague, J.A., Wang, C. & Topp, G.C. (1982). Estimating saturated hydraulic conductivity from soil morphology. *Soil Science Society of America Journal* **46**, 1239–1244.

Moen, J.E.T., Cornet, J.P. & Evers, C.W.A. (1986). Soil protection and remedial actions: criteria for decision making and standardization of requirements. In Asink, J.W. & van den Brink, W.J. (eds), *Contaminated Soil*, pp. 441–449. Martinus Nijhoff, Dordrecht.

Muckenhausen, E. (1962). *Enstehung, eigenschaften und systematik der Boden der Bundesrepublic, Deutschland*, DLG, Verlag, Frankfurt (Main).

Muckenhausen, E. (1965). The soil classification system of the Federal Republic of Germany. *Pédologie* numéro spécial 3, 57–89.

Mulcahy, M.J. & Humphries, A.W. (1967). Soil classification, soil surveys and land use. *Soils and Fertilizers* **30**, 1–8.

Oxenham, J.R. (1966). *Reclaiming Derelict Land.* Faber, London.

Polupan, N.I. (ed.) (1981). Polevoy opredelitel' pochv (*Field handbook for soil identification*). Kiev, Urozhay.

Protz, R. (1981). Soil properties important for various tropical crops: Pahang Tenggara master planning study. In *Soil Resource Inventories and Development Planning*, pp. 187–200. US Department of Agriculture, Soil Management Support Services, Technical Monograph No. 1, Washington.

Raeside, J.D. (1961). Letter to the editor. *Bulletin of the International Society of Soil Science* **19**, 20–21.

Ragg, J.M. & Clayden, B. (1973). *The Classification of some British Soils according to the Comprehensive System of the United States.* Soil Survey Technical Monograph No. 3, Harpenden.

Rozov, N.N. & Ivanova, E.N. (1967). Classification of the soils of the USSR. *Soviet Soil Science* **2**, 147–156.

Sencindiver, J.C. (1977) *Classification and Genesis of Minesoils.* PhD dissertation, West Virginia University, Morgantown (Diss. Abstr. AAD77-22746).

Short, J.R., Fanning, D.S., Foss, J.E. & Patterson, J.C. (1986a). Soils of the Mall in Washington, DC. II: Genesis, classification and mapping. *Soil Science Society of America Journal* **50**, 705–710.

Short, J.R., Fanning, D.S., McIntosh, M.S., Foss, J.E. & Patterson, J.C. (1986b). Soils of the Mall in Washington, DC. I: Statistical summary of properties. *Soil Science Society of America Journal* **50**, 699–705.

Soil Survey Staff. (1975). *Soil Taxonomy – a Basic System of Soil Classification for Making and Interpreting Soil Surveys.* US Department of Agriculture, Agricultural Handbook 436, Washington DC.

Soil Survey Staff. (1987). *Keys to Soil Taxonomy.* US Department of Agriculture, Soil Management Support Services Technical Monograph No. 6.

Soil Survey of England and Wales. (1984). *Soil Map of England and Wales*; in 6 sheets at 1:250 000 scale. Soil Survey of England and Wales, Harpenden.

Thompson, P.J., Jansen, I.J. & Hooks, C.L. (1987). Penetrometer resistance and bulk density as parameters for predicting root system performance in mine soils. *Soil Science Society of America Journal* **51**, 1288–1292.

Toy, T.J. & Shay, D. (1987). Comparison of some soil

properties on natural and reclaimed hillslopes. *Soil Science* **143**, 264–277.

Wallwork, K.L. (1974). *Derelict Land: Origins and Prospects of a Land-Use Problem*. David and Charles, Newton Abbot.

Wang, C. & Coote, D.R. (1981). *Sensitivity Classification of Agricultural Land to Long-term Acid Precipitation in Eastern Canada*. Land Resource Research Institute contribution No. 98, Ottawa, Ontario.

Wang, C., McKeague, J.A. & Topp, G.C. (1985). Comparison of estimated and measured horizontal Ksat values. *Canadian Journal of Soil Science* **65**, 707–715.

Watt, J.P.C., Griffiths, E., Cook, F.J. & Joe, E.N. (1986). Indicators and criteria for interpreting the hydraulic character of soils. In *Transactions of the 13th Congress of the International Society of Soil Science*, Hamburg, FGR **2**, 195–196.

Webster, R. (1968). Fundamental objections to the 7th approximation. *Journal of Soil Science* **19**, 354–366.

Webster, R. (1975). Sampling, classification and quality control. In Bie, S.W. (ed.), *Soil Information Systems*, pp. 65–72. Proceedings of the meeting of the International Society of Soil Science Working Group on Soil Information Systems, Centre for Agricultural Publishing and Documentation, Wageningen, The Netherlands.

Webster, R. (1977). *Quantitative and Numerical Methods in Soil Classification and Survey*, Clarendon Press, Oxford.

Yeterevskaya, L.V., Donchenko, M.T. & Lekhtsiyer, L.V. (1982). Systematics and classification of technogenic soils and recultivation in the USSR. *Abstracts of papers read at an All-Union Scientific and Technical Conference*, Part 2. Moscow.

3 Waste materials in urban soils

E.M. BRIDGES

The difficulties of classifying urban soils according to conventional systems of classification have been outlined in Chapter 2. Soil surveyors have usually avoided the issue by designating urban areas as 'built-up' and not attempted to classify and map their soils. Although considerable areas of the urban landscape have retained soils with a normal profile, many have been modified to a greater or lesser degree by man's activities. In some areas soils have been completely removed or buried beneath non-soil waste materials. This chapter is concerned with the waste materials which are found in urban soils.

An historical perspective

The sites of the earliest urban settlements in the Middle East such as Jerico (Jordan), Ur (Iraq) or Çantal Hüyük (Turkey) are all characterized by a raised mound or tell of up to 8–10 m above the surrounding countryside. Although construction of a mound may have been a response to attack or danger of flooding, archaeologists have found these mounds to be composed of the ruins of mud-brick houses and the rubbish thrown out by the inhabitants. Many present-day villages of the Middle East are underlain by similar mounds. In the case of Jerico it has been found that the tell contained the remains of many generations of structures which have yielded much interesting archaeological information extending back 12 000 years.

Until the fourteenth century most western European towns contained a mixture of rural and urban activities, with sufficient space within the walls to pasture animals and grow crops. These urban areas were more extensive than the tightly nucleated towns of the Middle East and so did not accumulate such impressive mounds of their own debris. It has been asserted that the levels of churchyards of the old city of Norwich are elevated because of the generations of people buried there.

Norwich may have been a special case, but mediaeval cities had houses built of mud and timber with roofs of thatch, materials which in the present

age would be called biodegradable. From their houses and premises the people 'threw their garbage, litter and offal into the street from doors and windows without regard to amenity or sanitation' (Trevelyan, 1947). Thus, urban areas up to the fourteenth century produced waste materials which were returned and incorporated into the soil, together with a few artefacts. The soils were relatively unadulterated by non-degradable wastes.

It was reported by Porteous (undated) that waste materials from towns were added to the soil of rural areas in an attempt to arrest the decline in fertility resulting from repeated cropping. Many of these substances arose from urban activities so they would inevitably have found their way into urban soils as well. The list included hog's hair, seal's hair, Fuller's earth, sugar baker's scum, soap boiler's ashes, grounds of malt vats, stagnant mud, fish broths, cartloads of starfish, sprats and seaweeds. The contents of middens and privies were also used to fertilize the soils around many towns and cities but many of these former rural areas around our towns have since been overwhelmed by suburban development in the last 30–40 years.

Waste materials since the industrial revolution

After the changes in industrial practices which took place about the beginning of the eighteenth century, collectively called the Industrial Revolution, the production of urban waste materials began to increase in amount and to change in content as industrial activities grew in importance. The process of industrial change has gathered momentum in the past three decades and many of the former heavy industries have closed and their sites have become disused or have been converted to alternative uses. In some cases sites have been used for many different successive purposes, some of which have resulted in soil contamination as waste materials have been added to the soil.

It is most convenient to survey the occurrence of waste materials according to the type of previous industrial occupancy of sites. A similar approach has been adopted by the Interdepartmental Committee on the Redevelopment of Contaminated Land (ICRCL, 1987) for problems of contamination.

Sources of waste materials

BUILDINGS

In the process of building, a certain amount of site disturbance is inevitable as excavations are required for foundations, cellars and services such as water and sewage. These disturbances apart, the soils of a building site may have been puddled and compacted. Debris from the building process may have been added to the soil but essentially it would have remained in place with a normal profile. However, there has always been a trade in topsoil, which unscrupulous builders sell before building commences, rather than stock-piling it so that it can be returned after the building operations are completed. Since the development of the bulldozer during the 1950s the soil cover of building sites has been increasingly disturbed. Instead of constructing houses in sympathy with the contour of a site, the bulldozer is used to construct a platform by cut and fill upon which the houses are subsequently built. In the process, the natural soils are greatly disturbed so that on some sloping sites it is possible to find buried profiles beneath bulldozed soil and parent material from further up slope.

From prehistoric times to the present day the most common waste materials in urban soils are from buildings. The accumulation of waste materials on prehistoric sites has already been mentioned, but with the expansion of urban areas on to greenfield sites during the last 50 years, changes have occurre in the amount and nature of waste materials. Building waste on a suburban area is generally restricted to broken brick, tile and glass with fragments of timber, piping, cable and insulation materials and discarded masses of cement, concrete and plaster. Most of these materials are more durable than the mud bricks and thatch of previous centuries and remain as discrete masses within the soil.

On derelict sites within older parts of a city, for example Tower Hamlets in London, little natural soil remains and the surface of the ground is covered with demolition debris from the buildings previously occupying the site. If such a site is redeveloped the presence in the ground of foundations, concrete floor slabs, underground cellars, drains and other pipework often causes difficulties. The sites of industrial premises may be more complex and the specialized nature of structures, often using reinforced concrete, can cause grave problems for the developer. Contamination is not normally present on former domestic housing sites but many industrial premises have this as an additional problem. Some of the problems encountered on the more intractable sites are discussed later in this chapter.

Clean building rubble is a useful material for hardcore but too many impurities detract from its value. Wood and other combustible materials are a common impurity, but a more intractable problem is the presence of plaster. Plaster is composed of calcium sulphate (gypsum) which in solution can cause corrosion of concrete structures. The Building Research Establishment has an advisory pamphlet on the requirements for concrete where sulphate-rich conditions are expected (BRE, 1981). Sulphate may attack concrete by internal crystal growth causing its disruption. Rusting of steel reinforcing rods also occurs and the whole structure is weakened. The problem is greatly enhanced when there is a high ground-water table in the soil which brings the building structure into the zone of the capilliary fringe. Examples of contamination by sulphate on sites in Lancashire are given by Barnes (1986). The ICRCL advises that sulphur up to 1000 mg kg^{-1} can be tolerated but where this amount is exceeded, further investigation should be undertaken. A trigger concentration of 250 mg kg^{-1} is set for the presence of sulphides and 1000 mg kg^{-1} for sulphates as extracted by solutions of hot dilute hydrochloric acid.

Bradshaw & Chadwick (1980) state that waste land in urban areas can be brought into productive cultivation by the use of a garden fork, 'Growmore' fertilizer, lime and organic matter. However, this optimistic approach overlooks some of the more intractable problems of urban soils, especially the presence of toxic metal elements. The nutrient status of clearance sites in Liverpool is given by the same authors (Table 3.1). The pH values which occur are usually over pH 7.0 and the calcium content is higher than in normal soils through the presence of mortar. The levels of nitrogen and phosphate are both low, but there are moderately good

Table 3.1 Nutrient levels in typical urban clearance areas, Liverpool (Bradshaw & Chadwick, 1980)

Site	pH	Ca (ppm)	N (%)	P (ppm)	K (ppm)
Grove Street	7.2	3830	0.05	36	220
Tennyson Street	6.7	1070	0.08	65	450
Brunswick Street	7.0	8830	0.05	19	290

supplies of potassium. Where disturbance is less, the soils of domestic gardens are often subject to application of excessive amounts of fertilizers, lime and pesticides. Disposal of fossil fuel residues, cinders, ash and soot and other household wastes also takes place, influencing urban soils to a greater or less extent.

Asbestos has been used widely for protection against fire as well as thermal and acoustic insulation in both industrial and domestic buildings. Recognition of its hazardous nature has led to a sharp decline in its use and it has been removed from many premises. Waste asbestos is most likely to be found in the soils of industrial premises, especially waste disposal sites, former railway land, shipbuilding and breakers' yards, power stations, scrapyards and former asbestos works. Where soils are wet there is relatively little danger of asbestos fibres being inhaled, but if the soil is disturbed in dry conditions then it can be hazardous. When derelict land is being redeveloped there is always some danger that both correctly disposed or fly-tipped asbestos will be uncovered. This is a likely occurrence where urban expansion has extended to include land upon which former waste dumping had occurred.

WASTE MATERIALS FROM METALLIFEROUS INDUSTRIES

At many places throughout the country, sites have been associated with the mining, smelting and manufacture of metals. Where mining has taken place, metallic elements will be present naturally in the soil in greater amounts (Davies, 1971; Davies & Roberts, 1978; Webb et al., 1978). However, ore concentration and smelting will result in further contamination of the environment and soils. In the Lower Swansea Valley, as a result of the former presence of copper, lead and zinc smelters, soils contained up to 900 ppm Cu, 9000 ppm Zn, 2700 ppm Pb, 27 ppm Ni and 27–90 ppm Cr; they were also badly eroded (Bridges, 1969, 1984). Trees have successfully been grown on these contaminated, eroded soils since 1962. In the restoration programme, some of the more highly toxic, metallic slags and soils polluted by leachate from the masses of slag have been sandwiched between layers of iron and steel wastes, the high pH of which minimizes the mobility of the toxic metallic cations. The final constructional surface has been covered by a layer of locally occurring glacial drift. These former smelter sites in the Lower Swansea Valley were grossly contaminated; some smelting residues remained which contained up to 11% Zn and adjacent contaminated soils had 168 ppm Cd, 3000 ppm Cu and 8000 ppm Pb (Davies, 1969). In such cases, where the toxic metals exceed the recommended trigger values (ICRCL, 1987), it is incumbent upon the local authority concerned to consider very carefully the future use of the land.

The processes of ore concentration can result in much fine-grained material. This results from the ore being ground to silt size before extraction with foam and chemicals. The waste 'slimes' are usually placed in lagoons until they dry out. Apart from the physical dangers of these lagoons where a thin crust develops over a still liquid mass, there is a dust problem from their surfaces. In urban areas associated with mining activities, former lagoons may be present and could contain residual amounts of toxic metals and possibly the separating agents which are also toxic. Red mud from aluminium production is another waste material present in some urban areas; disposal lagoons were associated with works at Burntisland in Scotland and Newport in South Wales.

Although the UK has a moist climate and soils are rarely dry for any length of time, there is still considerable movement by wind of finer particles as dust. Where the soils, or waste materials, contain toxic substances, hazards can result through dust inhalation or dust coating working surfaces where food is prepared. Atmospheric dust may be measured by filters or by absorption onto mosses. A survey of South Wales for the Welsh Office using moss bags found atmospheric contamination by toxic metals was widespread (Goodman & Roberts, 1971; Goodman & Smith, 1975). Other authors have found that household dusts contain appreciable amounts of toxic metals, but the link between these dusts and garden soils is often tenuous, there being too many alternative sources for metallic contaminants in the home environment (Welsh Office, 1983; Davies *et al.*, 1985; Thornton, 1986). The problem of toxic metals in the urban environment is dealt with more throughly by Thornton in Chapter 4.

WASTE MATERIAL FROM POWER STATIONS

The waste product from modern, coal-burning power stations is an ash which has 60% of the particles in the fine sand size (0.2–0.02 mm) and 40% in the silt size (0.02–0.002 mm). Clay-sized particles are virtually absent. This ash results from burning pulverized coal and is extracted from the flues by cyclones and electrostatic precipitators before the smoke and waste gases are passed up the chimney. Typically the dry ash contains about 48% silica, 26% aluminium, 10% iron oxides and relatively small amounts of other elements which are residual from the original coal. The levels of available boron (43 ppm), chromium (25 ppm) and molybdenum (5 ppm) are slightly above the levels found in natural soils but only the boron has been shown to be phytotoxic (Hodgson & Townsend, 1973). The physical condition of the ash makes it difficult to handle; it flows when aerated and is easily blown about by the wind. About 5% of the ash is composed of extremely light cenospheres, hollow spherical particles with a bulk density of about a quarter that of the rest of the ash. These small particles float on water and are a considerable dust problem when dry. To enable the ash to be handled more easily it is usual to add water to it to form a slurry and in this state it can be transported by pipeline to ponds where it is allowed to sediment from the water. The pH of fresh ash is about 9 and that of weathered ash about 8, but some ash with a greater content of soluble salts may have values of 11 to 12. Both the boron content and the soluble salt content are reduced by lagooning and this is recommended before use in land-restoration projects.

The ash reacts with lime to form a cement-like substance, consequently it is said to have pozzolanic properties, so it may be used as a building material.

When exposed to the atmosphere or in lagoons the ash also tends to form a cemented layer which limits permeability and penetration of plant roots. As the ash has no clay or organic matter it does not retain plant nutrients, so natural colonization by plants tends to be slow. Experiments with different depths of topsoil show that as little as 8 cm of soil enables adequate growth of grass and wheat, but wheat benefited by increasing the depth of soil to 30 cm (Townsend & Gilham, 1975).

Extensive land restoration has taken place south of Peterborough where the excavation of clays for brick-making has left many large quarries. Ash from three East Midlands power stations is taken by train in specially constructed wagons which can be pressurized to extract the ash which is added to water to make a slurry. This slurry is carried by pipeline to the quarry being infilled. After the sedimented ash has dried out, topsoil washed from sugar beet at a Peterborough factory is spread on the surface and good crops have been obtained (Barber, 1975). When it is desired to plant trees, it is necessary to excavate holes to 30–40 cm and backfill with soil material to give the trees a good chance of establishment.

Heavier particles of ash fall to the bottom of the furnace where they fuse together; this material can be ground to produce a sand for concrete or it can be discarded. Older power stations produce clinker, fused masses of glassy material, which has been extensively used as a fill material. No problems have been reported as a result of its use.

WASTE FROM CARBONIZATION PLANTS

The manufacture of town gas, tar and coke is achieved by the carbonization of coal. The by-product of ammoniacal liquor contains a wide range of derivatives used in the chemical and pharmaceutical industry. The process may be manipulated to maximize the desired end product. Gasworks were present in most towns and have now been closed or much reduced in operation with the access to natural gas from the North Sea. As these sites lie in urban areas they would command considerable value were it not for the contamination which is usually present. Many gasworks have been in existence for up to 100 years and spillages, leakage and deliberate dumping of wastes took place within the works.

Clinker and spent oxides were often disposed of on site. The glassy clinker is relatively harmless, but the spent oxides are strongly acidic and contain ferric ferrocyanide. These oxides were used to remove impurities from the gas including hydrogen sulphide, cyanide, ammonia and other trace chemicals. With repeated use, the iron oxides formed iron sulphides and sulphur and when the sulphur content reached 50–60% the oxides were discarded. The coal tar contains many toxic compounds including benzene, toluene, xylene, ethyl benzene, styrene, phenol, cresols, xylenols, polyhydric phenols, naphtha, naphthalenes, acenaphthelene, fluorene, diphenyl oxide, anthracene, phenanthrene, carbazole, tar bases and pitch (ICRCL, 1979). Some of these carcinogens have had 100 years to seep through the soil and contamination may be found to depths of 10 m or more, and seepage downslope may extend well beyond the gaswork's fence. The ammoniacal liquor similarly contains many hazardous compounds which are valuable in the by-products industry.

Land released from the Beckton Gasworks in London was so badly contaminated that it had to be covered by 1.2 m of London Clay covered with 0.3 m of topsoil. Beneath the London Clay the wastes were covered by a 225 mm thick layer of gravel to act as a drainage channel to convey water away from the wastes and to act as a venting channel for gases emanating from them.

OIL SPILLS AND OILY WASTES

Oil spills at sea have caught the headlines, but there are also numerous oil spills on land which have been awarded less glamorous treatment by the media. Oil spills may be accidental or deliberate and are of considerable concern to conservationists.

Evidence is emerging of several examples of accidental oil or petrol spills through leaking underground tanks or pipes at garages and filling stations. Accidents to road or rail tankers have also resulted in significant quantities of oil or petrol entering soils. When petrol is spilt on soil it rapidly flows down soil pores and may, if the substrate is porous, reach the water table. At the water table the petrol will remain on the surface of the water and most of it can be reclaimed by sinking a bore hole into the site and drawing up a mixture of petrol and water.

Crude oil and heavier lubricating oils infiltrate more slowly into the soil, so it is preferable quickly to scrape up as much as possible before it penetrates deeply. Any residues which remain should be cultivated to aid volatilization and reduce anaerobic conditions. Burning the oil is not recommended (CONCAWE, 1983).

Where petroleum and oil spills occur, crops are tainted and soils may become anaerobic if completely saturated. Liming, and the addition of nitrogen and phosphatic fertilizers, is reported to encourage the microbial population to break down the hydrocarbons. Breakdown of soil structure has been observed initially, but reports suggest that there is a subsequent improvement as the bacteria break down and incorporate the residues.

Waste oils are a problem in that they often contain contaminants from the processes in which they have been used as lubricants. At Times Beach, Missouri, heavy oils containing dioxins were liberally sprayed on dirt roads to keep down the dust. The waste oils were derived from a factory making the defoliant 'agent orange'; the land was so badly contaminated with dioxins that the Federal Government has purchased the whole area.

CHEMICAL WASTE MATERIALS

The problems of chemical waste materials in the soils of urban areas has been brought to the attention of the general public through sites such as Love Canal, near Niagara in the USA and Lekerkerke in the Netherlands. Numerous other sites have been documented in England, Wales, Germany, Canada and other countries (Smith, 1982).

The Love Canal was an attempt by William Love to link the upper and lower sections of the Niagara river, above and below the Falls, by a canal in 1896. The project was abandoned, leaving a 3000 m trench which was eventually purchased by a chemical company in 1942 for use as a dump. By 1953 some 19 000 t of solid and liquid chemical waste had been deposited and buried. The site was covered and eventually a school and houses were built upon it. By 1977 the buried chemicals had begun to migrate as the drums containing them began to disintegrate, some were beginning to seep into basements and people were beginning to experience unexplained illnesses. In 1978 the school was closed and 235 families who lived within 120 m of the former canal were evacuated. The following year families with children under two and with pregnant mothers were moved from the southern part of the site and the remaining families were advised not to initiate pregnancies (Paigen *et al.*, 1985). Two hundred and forty eight different chemicals were identified from the Love Canal site including benzene, carbon tetrachloride, vinyl chloride, dichloroethane, hexachlorobenzene, hexachlorocyclohexane, lindane, polychlorinated biphenyls, trichlorophenols, tetrachlorodibenzine-*p*-dioxin, toluene and xylene. Altogether, there were 34 neurotoxins, four pulmonary toxins, 20 hepatoxins, 15 renal toxins, 34 carcinogens, 18 teratogens and 30 foetotoxins or embryotoxins. As a result of the presence of these substances greater than average health problems were experienced by residents including asthma, urinary tract problems, strictures, renal failures, central nervous system problems, miscarriages, stillbirths and birth defects. Other problems included seizures, learning problems, hyperactivity, eye irritations, skin rashes, intestinal problems and incontinence. The cost in terms of human misery was immense and the Love Canal disaster was largely responsible for persuading the US Government to release funds to implement the Comprehensive Environmental Response, Compensation and Liability Act of 1980 (CERCLA), better known as Superfund.

At Lekerkerke in the Netherlands, houses were built upon an old dump containing waste materials from a dyestuffs factory. The factory which had produced these wastes had been demolished and was so contaminated that is was dumped in the Atlantic, but the wastes remained buried. They included drums of toluene, xylene, and compounds of lead, cadmium and zinc. Three hundred houses had to be evacuated and 1.5 million tonnes of polluted soil removed from beneath the houses before the inhabitants could return safely (Schuuring, 1981). In another town, Gouderak, houses were built on wastes from oil refining which had been dumped between embankments on the river forelands. People who lived in the houses began to have medical problems which were eventually traced to the wastes beneath the houses. The occupants were evacuated and the houses were demolished (Bins-Hoefnagels & Molenkamp, 1986).

Clearly, these are some of the worst examples, but throughout most industrialized countries, problems of soil contamination by waste chemicals occur to a greater or lesser extent. However, in Britain, as old dumps are cleared the position should gradually improve as the Disposal of Poisonous Waste Act (1972) and the Control of Pollution (Special Waste) regulations of 1974 require that toxic wastes are to be disposed of only at specially licenced sites. On the basis of relatively short-term research, Britain permits co-disposal of toxic chemical wastes with domestic waste, even though an EEC directive has suggested that co-disposal of chemical and domestic wastes is undesirable (Sumner Report, 1978; Cope *et al.*, 1983).

One of the first industrial chemical processes to be developed was the manufacture of washing soda. In the LeBlanc process, sodium carbonate was produced with alkaline wastes including calcium hydroxide and calcium sulphide which were usually dumped beside the works. The process became obsolete in the second decade of the present century and the works dismantled, but the dumps remain at several locations, for example on Tyneside, at Widnes, St Helen's and at Flint in North Wales.

Sodium carbonate was subsequently produced by the Solvay process in which brine is saturated with ammonia and then allowed to trickle down a carbonating tower against a current of carbon dioxide. The sludge which reaches the bottom of the tower is then heated to drive off the moisture. The resulting sodium carbonate is heated with calcium hydroxide to produce sodium hydroxide with calcium carbonate as a waste product which is dumped. Although strongly alkaline, these wastes are not toxic and so have been used as cover materials for strongly acidic or metallic wastes in restoration programmes on other contaminated sites.

Calcium sulphate, gypsum, is produced as a waste product from the manufacture of orthophosphoric and hydrofluoric acids from phosphate rock. Although 2 million tonnes of phosphogypsum per annum are discharged into the estuaries of the Severn and Trent, some is disposed of in lagoons on land but the leachate contains flouride in sufficient quantities to pollute watercourses (Gutt *et al.*, 1974).

For many years chromium salts have been used as pigments, in the tanning of leather and in recent years the metal chromium has become important in the manufacture of special steels. Currently, chro-

Miata

mium is smelted from the ore chromite in an electric furnace but formerly extraction of the metal was achieved by smelting with sodium carbonate or with a mixture of sodium and calcium carbonate. The chromite was oxidized to sodium chromate and subsequently removed from the melt by leaching. The remaining waste material, mainly calcium salts, was dumped. Even after two leachings, the waste still contained up to 2% water-soluble chromate. Problems have been encountered with the toxicity of the leachate from these dumps and their surface is unstable because sodium sulphate contained in the waste is water soluble and leads to subsidence (Gemmell, 1973).

Many different solvents are used in the chemical, paints and dyestuffs industries. The ground below former works using solvents may contain quantities of solvent as a result of accidental spillage, or deliberate dumping. It is not necessary to detail all of the chemicals used but it can be summarized by stating that the aliphatic hydrocarbons are not excessively toxic; aromatic and cyclic hydrocarbons are more dangerous, especially benzene; and the alcohols, ketones and esters are irritants rather than being toxic. The Waste Management Papers Nos 14 and 23 produced by the Department of the Environment, (DoE, 1977, 1981), list most of the solvents used by industry and indicate their toxicity, corrosivity and carcinogenic potential. The management techniques to be used for their safe disposal are also given. Where solvents are found in the soil or in tanks and pipework buried in the ground, they should be treated carefully and specialist contractors employed to remove them. Under no circumstances should these liquids be allowed to leak out to cause further contamination.

The manufacture of pesticides, including insecticides, fungicides and herbicides may lead to dangerous levels of waste chemicals in the soil. Any site where these chemicals were manufactured, handled or their residues dumped is of potential danger to subsequent occupants. The problems of a waste dump in California which contained 21 pesticide compounds including dibromochloropropane are described by Dahl (1986). The problem with these and similar substances is that they are either neurotoxins, carcinogens or teratogenic compounds.

Industrial concerns which have been responsible for leakages of solvents have included contemporary industries manufacturing silicon chips. In the USA, tanks of solvents used in the manufacture of silicon chips leaked to give several hundred parts per billion of these fluids in drinking water in Silicon Valley. A report in the *New Scientist* on 21 November, 1985, stated that 10% of Britain's aquifers were polluted by solvents according to a report submitted to the Department of the Environment. Some of the most common solvents found in waters are trichloroethylene and chloroethylene, which are used in paper-making, metal plating, electrical engineering, laundries, dry cleaners and as degreasing agents. In Scotland, 61 aquifers used for drinking water out of 168 sampled contained traces of trichloroethylene, many of which were at levels above the WHO guideline of 30 ppb; one had 204 ppb. Another example of solvent pollution occurred in Suffolk in 1981 where water containing 286 ppb trichloroethylene was traced to a chemical sump at an air force base at Mildenhall.

Land formerly occupied by munitions factories has been found to possess many problems because waste materials remain in the soil. Development of the former Woolwich Arsenal at Thamesmead revealed heavy contamination which brought building work to a halt whilst further investigations were made (Lowe, 1980). In addition to the possibility of hazardous materials from the explosives, the arsenal had its own gasworks and facilities for electricity generation. Metallurgical work in the construction of armaments left a legacy of metallic contamination, with copper, nickel, cobalt, strontium, barium and mercury salts being present in the soils and masses of metallurgical slag scattered around the site where it was used to raise the level of roads above the marshlands. However, one of the major problems at Thamesmead was tanks of phenolic liquids buried beneath the site of the former gasworks. Numerous massive concrete bunkers, surrounded by water, which were used for storage and protection are still present and are to be retained for leisure uses. Similar bunkers occur on the site of a former ordnance factory, now a country park, at Pembrey in Wales. Establishment of plants on small areas of soil polluted by burning-off explosives has proved difficult, but the contaminated soil has been removed and replaced with clean sand (Bridges, 1988).

In recent years concern has been expressed about

the chemical compounds known as dioxins. These substances include some of the most poisonous chemicals produced. They occur as contaminants in the manufacture of several organic pesticides and also are produced by incinerators burning plastics at low temperatures. Dioxin is held in the human body longer than in other animals and a recommended acceptable daily intake should not be more than 0.1 pg (Mackenzie, 1985). Examples of contamination of urban soils by dioxins include the 1800 hectares around Seveso, Italy, where, following an accident at a chemical works, 736 people had to be evacuated from their homes, 236 of them permanently. Contaminated soil was removed from the area and sealed in drums which subsequently disappeared, eventually re-appearing in a disused abbatoir in France. Friends of the Earth claim that two times the level of dioxin found at Seveso occurs in some Scottish soils including 2378 TCDD emanating from a waste disposal facility. Destruction of pharmaceutical wastes in Ireland at an insufficiently high temperature in an incinerator resulted in the headline in the *New Scientist*: 'Irish incinerator sprays farm with toxins'. Burning of polychlorinated biphenyls following the explosion of a transformer in Rheims in France also resulted in dioxins being released into the surrounding environment.

It has been claimed that smectite clays can be used to detoxify dioxins. Highfield (1985) reported that smectite clays treated with copper solution and then freeze-dried forming a copper smectite can be used as an electron acceptor from the dioxin making it ready for detoxification by polymerization or reaction with other chemicals.

ORGANIC WASTES IN SOILS

In an attempt to delay the inevitable impoverishment of land which occurs during cropping, farmyard and stable manure have been returned to the land as long as settled agriculture has existed. Many town gardens also have been enriched by digging-in rotted farmyard manure. Where this has occurred on a regular basis in old gardens, the topsoil has a dark colour, and a high phosphate and organic matter content. The average composition of farmyard manure is 0.6% N, 0.1% P and 0.5% K with about 24% organic matter (Bradshaw & Chadwick, 1980).

In the past 40 years intensive rearing of chickens in broiler houses or battery units has given rise to significant quantities of poultry manure. Bird droppings contain uric acid and ammonium compounds which must be broken down by composting before use on the land. Once this has been accomplished, poultry manure is richer in nutrients than farmyard manure. The average composition is given as 2.3% N, 0.9% P and 1.6% K with 68% organic matter.

Sewage sludge is another organic waste product, the production of which has increased during the past 40 years. Typically, it contains 2.0% N, 0.3% P, and 0.2% K and is 45% organic matter. Sewage sludge has been widely used on urban gardens and amenity sites to increase the organic matter content of soils and to act as a source of nutrients. However, it cannot be used in unlimited amounts as it contains toxic metals. Sewage disposal on agricultural land may mean that the soils become polluted by toxic metals. Published figures are available for the former sewage farm at Beaumont Leys, Leicester, where after 74 years of sewage sludge disposal, the soils contained totals of up to 3000 ppm zinc, 1400 ppm copper, 385 ppm nickel, 240 ppm lead, 60 ppm cadmium, 2000 ppm chromium, and arsenic up to 60 ppm. Extractable amounts were between one quarter and one third of these amounts (Heeps & Pike, 1980). As a result of concern about the levels of toxic metals contained in sewage sludge, experiments by the Ministry of Agriculture led to the adoption of a zinc factor which reflected the combined effects of zinc, copper and nickel. Chumbley (1971) gives a table of permitted applications of sewage sludge over a 30-year period which could be safe assuming the soil pH is about 6.5.

Mushroom compost originates as animal wastes and straw which have been well decomposed and limed; once used for mushrooms, it can become a useful organic addition to soils. It could be used on urban ground to increase rapidly the organic and nutrient content of amenity soils. Mushroom compost contains 2.8% N, 0.2% P and 0.8% K, and is 95% organic matter.

In older gardens, many urban soils formerly will have received human wastes from the contents of the privy but water-carried sewage systems now serve most areas of Britain and so this practice is declining. However in isolated places, and in many other countries, disposal of sewage from septic tanks occurs on site and the garden soil is used to absorb waste waters from the septic tank. For the successful operation of these units the permeability of the soil should be capable of dealing with the expected amount of effluent.

Suitably-treated domestic wastes may be used as organic fertilizer in certain circumstances. The problems of landfill will be dealt with in a subsequent section, but where domestic wastes have been suitably sorted, pulverized and composted, they have been used as a soil conditioner or top-dressing. However, the resulting compost has a high C:N ratio (about 30:1), so its fertilizer value is minimal (Porteous, 1975). After incineration, the residues of domestic wastes are normally landfilled or otherwise used as aggregate. Incineration concentrates the toxic metal residues, so these may be unsuitable for use as soil or subsoil in reclamation schemes.

Waste materials from forest products are extensively used on urban soils as mulches to suppress weed growth. Lutz & Chandler (1947) have shown that the ash content of tree species is different for stem wood, branches, bark and leaves; the chemical composition of these is also different for different species. Softwood barks decomposed more slowly than the bark of hardwood species but addition of nitrogen fertilizer caused a marked increase in the rate of decomposition of hardwood bark whereas only some species of softwood were affected. The presence of bark from certain trees caused a significant decrease in the growth of test crops of peas and some were very injurious, so caution is advisable when adding these residues to soils (McCalla *et al.*, 1977).

WASTE MATERIALS ON TRANSPORT LAND

Abandonment of the canal system and contraction of the railway network has left considerable areas of derelict land in urban areas. Where wharves, stations and goods yards occur in urban locations, this land can be extremely valuable. Linear sections of canal bed or rail track may be made into walkways or cycle paths, but embankments, tunnels and cuttings are not so easily redeveloped.

Land associated with canals may have underground passages through which water was transferred in and out of the locks. With the passage of time, many canals have been completely silted up and the sludges on the puddled floor of the canal often contain toxic metals and other debris dropped into the water. In Lancashire, enhanced levels of mercury have been found in canal sediments adjacent to a chlor-alkali works and an examination of sediment in the Neath Canal found 152 ppm copper, 441 ppm lead, 953 ppm zinc and 3 ppm cadmium. Similar sediments from harbours look suitable for topsoil dressings on reclaimed areas; however, before they are used a careful check should be made of toxic metal concentrations (Bradley & Rimmer, 1988). Adjacent to canals, soils are often poorly drained and production of methane may result from anaerobic breakdown of any organic matter present. Similar problems are experienced in the redevelopment of dockland areas.

On former railway land the lines and sleepers are normally removed for their scrap value, but usually the ballast remains on site. Minor contamination may occur along the line of the railway but it is in the former railway yards that unpleasant wastes may have been added to the soil. A problem of combustibility exists at engine depots where boilers were raked-out and the cinders and waste coal thrown away. Former railway land at Swindon suffered a subterranean fire which threatened the Wiltshire Technical College and a hospital laundry in 1973. Examination of former sidings indicated the presence of toxic metals in excess of recommended safety threshold values. Destruction of electrical equipment may have left residues of polychlorinated biphenyls and if former rolling stock has been dismantled there may be asbestos dumps on railway land. Oils, degreasing fluids, paint residues and solvents may also be found as contaminants.

WASTE MATERIALS IN SCRAPYARD SOILS

The sites of former scrapyards are proving to be some of the most highly polluted soils of urban areas. Dismantling of obsolete machinery, such as vehicles, electrical equipments and other consumer goods, has left a residue of metals, asbestos, oils and other substances which are phytotoxic, a danger to human health or aggressive to building structures which are placed upon the site. Other hazards which may be present on former scrapyard sites include problems of combustibility, and emissions of toxic or flammable gases (Bridges, 1987).

Sites which have been examined have shown the

likelihood of arsenic, antimony, barium, cadmium, chromium, copper, iron, lead, manganese, mercury, nickel, tin and zinc being present in sufficient quantities to be a potential danger to plant, animal and human life. Chemical substances which may be present on a scrapyard site may include many of the substances mentioned in the section about chemical wastes. These may include cyanides, chlorides, fluorides, phosphates, sulphides and sulphates as well as acids and alkalis used for degreasing or cleaning metal objects. Often organic solvents are responsible for contamination at these sites, but one of the more serious liquid wastes arising from the break-up of electrical equipment are the polychlorinated biphenyls (PCBs) used as dielectric fluids in transformers. At one site the presence of these substances in the soil increased the estimated cost of reclamation by £250 000 because the saturated soil had to be removed and disposed of before other work could begin (ICRCL, 1980).

The presence of soils saturated with oils in scrapyards leads to problems of combustibility and anaerobicity which in turn may result in gaseous emissions which are discussed in connection with wastes associated with landfill sites. If services for new buildings are laid in these contaminated soils they are likely to be rapidly corroded and weakened; plastic water pipes are said to be susceptible to damage by oily substances and phenolic liquids. If concrete foundations are placed in materials contaminated by strongly acidic substances, they may be weakened to the detriment of the buildings they support. Failure of structures has been observed by some local authorities on contaminated soils where precautionary measures were not taken.

WASTE DISPOSAL SITES

Although the bulk of domestic waste deposited in landfill is innocuous, it does contain significant quantities of toxic elements including cadmium, copper, lead, mercury, nickel, copper and zinc. These metals become more soluble and mobile in the anaerobic conditions of a landfill. Also, co-disposal of toxic or otherwise hazardous materials often has occurred either with or without official knowledge. This means that redevelopment of landfill sites will always be difficult, so they are probably best avoided, especially as use of such sites greatly increases foundation costs. Other difficulties include the presence of asbestos, pathogenic organisms, the danger of gas and leachate generation and the problems of combustibility and smells.

When tipping has finished, landfill sites are usually covered by a thin layer of soil material so plant growth generally is quite good except where phytotoxic elements are present or where the voids in the substrate become filled with gases. In recent times more comprehensive cover systems have been devised with several layers, each performing a specific purpose. All designs for cover systems incorporate a barrier layer, which may be of clay, limestone, or gravel or plastic which physically or chemically isolates the hazardous wastes.

For example, a landfill site in the London area, destined for redevelopment was found to have a heavy metal toxicity problem, so a cover system was designed to isolate the waste and the metals from the surrounding environment. First the site was covered with 150 mm coarse hardcore to stop capillary rise and to form a drainage layer. Next 25 mm of fine porous material was placed over the hardcore to act as a moisture reservoir and to stop the final layer of 150 mm of topsoil from falling down between the coarse gravel fragments. Where trees or houses were anticipated a depth of 550 mm of clean topsoil was specified. It is not known whether this proposed redevelopment has taken place.

Plastic sheeting has been used on an experimental basis in some countries to stop water from getting into the wastes and generating further leachate. At a site in the USA, the landfill surface was first graded to gentle slopes to shed water from the surface and sides. A 100 mm cover of sand and gravel was placed over the refuse to act as a drainage layer and to protect the plastic barrier from sharp objects in the waste. Further sand and gravel was then placed over the plastic barrier to a depth of 450 mm. Decomposed leaf litter and sewage sludge were incorporated into the surface and the site sown with grasses. A cut-off ditch was dug to encircle the site to divert surface run-off (Sanning, 1980).

WASTE MATERIALS OF SEVERELY DISTURBED AREAS

Virtually all of the areas mined for coal or ironstone by the opencast method were rural and were restored to some form of agriculture or forestry. However, some were closely associated with built-up areas on the coalfields, and colliery wastes have been extensively used as fill for roadworks and other civil engineering projects, many of which are in urban areas. Consequently, it is appropriate to comment upon these materials in the context of waste materials in urban soils.

A survey of waste materials in 1974 revealed that about 2000 million tonnes of colliery waste existed in various parts of the UK (Gutt *et al.*, 1974). This material is rejected either at the pithead or washery and consists of rocks other than coal which are brought to the surface. In the past when miners hewed coal by hand, there was little waste other than when sinking the shaft or driving tunnels through unproductive rock. Modern mechanical methods of mining have increased the proportion of waste which subsequently must be separated from the coal and disposed of in a waste heap.

Colliery waste comprises sandstone, mudstone, siltstone, shale and seat earths which are associated with the coal seams. The Coal Measures are rocks laid down in a deltaic sedimentary environment which alternated between freshwater and saltwater conditions. Either changes in sea level or pulsed tectonic events resulted in the accumulation of cyclical deposits of coal, sandstone and shale, which form the strata of the Coal Measures. The seat earths were rudimentary soils in which the plants of the Carboniferous deltaic swamps grew. Some of these deposits were leached and kaolinitic clays developed, but poor drainage led to reducing conditions and accumulation of pyrites. These pyritic deposits are often found in association with black shales in the geological succession and when weathered the pyrites oxidizes to give strongly acidic conditions, causing acid drainage waters and inhibition of plant growth in the restored soils of opencast sites and amenity areas.

Some Coal Measures deposits have retained the salinity associated with the environment in which they were deposited; sodium chloride, potassium chloride, magnesium chloride, magnesium sulphate and calcium sulphate may be present. When the rock material is exposed to the elements it rapidly releases these soluble salts which pass into the soil solution. Although eventually leached in the humid British environment, salinity has caused problems of plant establishment on restored sites in Northumberland and Durham. Where old colliery waste heaps have been regraded, unweathered shales have been exposed and salinity problems have re-emerged.

The general nutrient status of colliery waste for plant growth as revealed by soil analysis and pot experiments indicates that nitrogen and phosphate are lacking and that only potassium is present in sufficient quantity for adequate plant growth. Generally, micronutrients are available in sufficient quantities for plant growth. Applications of fertilizer are necessary to obtain satisfactory plant growth from colliery wastes, particularly nitrogen which should be given in several small applications rather than one large one. As there is little or no available phosphate, enough should be provided for plant growth and to counteract losses through fixation. The phosphate should be cultivated deeply into the ripped surface to encourage plant roots to penetrate deep enough to obtain sufficient moisture. Some shale wastes from Scottish collieries are neutral or alkaline but many colliery wastes in England and Wales are acidic and liming is necessary.

Colliery waste has been widely used on many urban civil engineering schemes throughout Britain. Hitherto, burnt spoil was preferred as there was less likelihood of it igniting and burning below the ground surface. However, with suitable compaction, the unburnt shale may be used and forms the subsoil of many amenity soils on engineering projects. On the coalfields many areas have been disturbed by opencast mining and their soils 'restored'. These areas have almost certainly been avoided for urban development because of the expense of adequate foundations for buildings, but where they occur in urban areas, opencast sites may be used for public open space, sports fields and golf courses.

However, colliery shale is not soil and the time taken to convert mixed overburden into soil is lengthy and a natural soil profile is slow to develop. A significant feature of soil, as compared with parent material, is the development of soil structure. As Richardson (1975) explains, the development of soil structure is assisted by the supply of polysaccharide gums which bind soil particles and stabilize the structure. These gums are produced by soil microorganisms, the activity of which can be used as a measure of the fertility of these reconstructed soils.

Parts of Oxfordshire, Northamptonshire, Leicestershire and Lincolnshire have been excavated for the Jurassic iron ores which occurred at shallow depth. These areas were almost all rural and although the landscape has been restored to agriculture, it may be 2–3 m below the original level. Fortunately, limestone is a beneficial material, raising the pH of leached soils and helping to promote vigorous faunal activity and to improve soil structure. It is a material used extensively in civil engineering works and so quantities of it will also occur in urban soils. As on the restored opencast coal sites, the land is unlikely to be used for building as foundation costs are greatly increased. However experiments into the most effective methods of stabilizing the disturbed rock for foundations have taken place at Corby in Northamptonshire (Rimmer, 1976). Overburden from the Carboniferous limestones of Wales, referred to as 'scalpings', has been used as cover material for landfill (Bridges *et al.*, 1986).

Products generated from waste disposal

WASTE GASES IN THE SOIL

The air which occupies the pore spaces of freely drained soils differs in two major respects from the atmosphere; its content of carbon dioxide is approximately eight times as great and it is saturated with water vapour. In natural, poorly drained conditions, water fills the soil pores and anaerobic decomposition of organic matter leads to methane production and the phenomenon Will-o'-the-Wisp. The microbial decay of landfill wastes containing organic materials produces a mixture of gases of which methane is the major component, but it also contains carbon dioxide, hydrogen sulphide, hydrogen and nitrogen. Minor quantities of other substances, such as mercaptans, are responsible for the noxious odours. It has been estimated that most of the methane generation takes place in the first 5–10 years of a landfill, but it may persist for up to 30 years. If disturbance occurs, or the level of the water table in the waste is changed, the process of gas generation may be reactivated.

Generally, these landfill gases are allowed to escape into the atmosphere. As the gas takes particular pathways through the wastes, it tends to damage patches of vegetation growing on the cover materials. Methane is a highly mobile gas which can migrate from the landfill into the pore spaces of undisturbed soils around the site. Numerous examples exist of methane accumulation in cellars and cupboards of houses, which when ignited by accident, have exploded causing damage and loss of life. Where gas migration is troublesome, it may be necessary to surround a landfill site with a trench, lined on the outside with an impermeable membrane and back-filled with coarse gravel to allow venting of the gases. Where development has taken place on former dockland sites, which are also liable to produce methane, it has proved necessary to incorporate venting systems in the building design (Carpenter, 1986).

Other volatile substances which have caused problems in soils have leaked from underground storage tanks or have emanated as leachate from chemical waste dumps or spills. James *et al.* (1985) discuss the problems of volatilization of chemical wastes and the presence of these substances in soils on sites which require remedial action. They list the more common substances found on former dump sites in the USA, with their flash points, boiling points, molecular weights and various threshold limits. Studies have shown that the main volatile organic compounds found in soils adjacent to hazardous waste landfills in the USA were toluene, benzene, phenol, ethyl-benzene, naphthalene, vinyl chloride, methylene chloride, chloroethane, trichloroethylene, chlorotoluene, tetrachloroethane, chlorobenzene, xylene, trichloroethane, chloroform and methane. Where waste gases have filled the pore spaces of soils, the most common visual effect is damage to, or the absence of, vegetation as the presence of the gas inhibits the functions of the roots and the aerobic microbial population.

WASTE LIQUIDS (LEACHATES) IN SOILS

In any environment where the precipitation exceeds evaporation the rainfall becomes a critical factor in leachate generation from landfills and other sites where toxic waste materials have been dumped. In this context, leachate is defined as liquid emanating from a contaminated site. It may be liquid products of decomposition, liquid wastes, percolated precipitation, or groundwater flowing through the site.

Other factors which must be considered are the local topography, the nature of the soil and the level of the groundwater table. Where a permeable rock underlies a landfill, leachate is filtered and made innocuous as it percolates downwards. However, if the groundwater table is close to the base of the landfill, toxic substances may quickly appear in the groundwater. A landfill in an impermeable rock will gradually become saturated until it overflows and the resulting leachate contaminates adjoining land and watercourses. With saturation and anaerobic conditions, solubility of many substances increases and the concentration in the leachate is greater.

Several examples have been documented where leachates from toxic dumps have polluted soils in urban or peri-urban areas. The problems associated with the Love Canal chemical dump site at Niagara have been mentioned already. Another example from the USA is the Picillo site at Coventry, Rhode Island where leaking drums containing waste toluene, xylene, trichloroethane and trichloroethylene were polluting the area surrounding the dump. The drums had to be removed. In London, seepages of tar and creosote from lagoons at the East Greenwich gasworks have penetrated through to the Blackwall Tunnel. Steif (1984) provides examples from Germany. Pollution by liquids containing arsenical compounds near Cologne, tetrachloroethylene at a site in Baden Wurttemburg, aromatic and aliphatic hydrocarbons at Frankenthal, Rhineland Pfalz and insecticide wastes at Gendorf in Bavaria, are described. At considerable expense these pollutants had to be contained or removed from the soil. Accidental spillages also contribute to the pollution of urban soils; a fractured pipe at a works in Regina, Canada allowed up to 21 000 l of a liquid containing PCBs and TCBs to escape. Fortunately, most was collected and removed following the spillage. Leaking petrol tanks at service stations are another hazard.

Standards for risk assessment

Inert wastes may or may not be a physical problem if they occur in urban soils, but where toxic wastes have been left behind by previous occupants of the land, the environmental scientist is presented with three extremely delicate problems: what is hazardous, what is the risk and what should be done about

it? Unfortunately the relationship between the presence of toxic materials in the soil and the plants, animals and human population is a tenuous one and the question is not answered simply. Consequently, countries aware of the problems of wastes in the soil have been trying to work towards a definition of the concentration of contaminants in soils which can be considered to be safe, where a possibility of danger exists, and where removal of the contaminant is considered to be necessary. Reference books such as Sax and Lewis (1989) indicate the dangerous properties of most industrial materials.

In the UK, the Interdepartmental Committee on the Redevelopment of Contaminated Land (ICRCL, 1987) has produced guidelines for several contaminants. Currently, 'trigger concentrations' below which soil is considered to be unpolluted are given, as well as 'action values', above which remedial action is required. The proposed values for non-metallic contaminants associated with coal carbonization plants are reproduced in Table 3.2. The values for concentrations should be compared with the results from individual 'spot' samples based on an adequate site investigation. Between the trigger concentration and the action value any decision about remedial measures is dependent upon the experience and skill of the scientist in charge of the redevelopment. In this way the UK system tries to vary the standard applied according to the proposed use of the land after rehabilitation. This is in contrast to the Dutch approach which endeavours to achieve a complete clean-up of contaminated land (Beckett & Simms, 1986; Finnecy, 1987).

The tentative trigger values put forward by the UK Department of the Environment should only be used within the limits prescribed by the guidelines (ICRCL, 1987). For example, the threshold trigger concentration for polyaromatic hydrocarbons is $50\,mg\,kg^{-1}$ in domestic gardens but is allowed to rise to $1000\,mg\,kg^{-1}$ for public open space and sportsfields; similarly the trigger concentration for free cyanide is set at $25\,mg\,kg^{-1}$ for domestic gardens and $100\,mg\,kg^{-1}$ for buildings and hard (concrete or asphalt) ground covers.

The Dutch Government have proposed a reference value (A) indicating the concentration present in an uncontaminated soil, a second value (B) indicating the need for further investigation and a third figure (C) above which a clean-up exercise is

Table 3.2 Tentative 'trigger concentrations' for contaminants associated with former coal carbonization sites (ICRCL, 1987)

Contaminants	Proposed Uses	Trigger Concentrations Threshold	Action (mg kg^{-1} air-dried soil)
Polyaromatic hydrocarbons[1,2]	Domestic gardens, allotments, play areas	50	500
	Landscaped areas, buildings, hard cover	1 000	10 000
Phenols	Domestic gardens, allotments	5	200
	Landscaped areas, buildings, hard cover	5	1 000
Free cyanide	Domestic gardens, allotments landscaped areas	25	500
	Buildings, hard cover	100	500
Complex cyanides	Domestic gardens, allotments	250	1 000
	Landscaped areas	250	5 000
	Buildings, hard cover	250	NL
Thiocyanate[2]	All proposed uses	50	NL
Sulphate	Domestic gardens, allotments, landscaped areas	2 000	10 000
	Buildings[3]	2 000(3)	50 000(3)
	Hard cover	2 000	NL
Sulphide	All proposed uses	250	1 000
Sulphur	All proposed uses	5 000	20 000
Acidity (pH less than)	Domestic gardens, allotments, landscaped areas	pH5	pH3
	Buildings, hard cover	NL	NL

[1] Used here as a marker for coal tar, for analytical reasons. See 'Problems Arising from the Redevelopment of Gasworks and Similar Sites' Annex A1.
[2] See 'Problems Arising from the Redevelopment of Gasworks and Similar Sites' for details of analytical methods.
[3] See also BRE Digest 250 (1981): Concrete in sulphate-bearing soils and ground water.
NL: No limit set as the contaminant does not pose a particular hazard for this use.

Conditions
1 This table is invalid if reproduced without the conditions and footnotes.
2 All values are for concentrations determined on 'spot' samples based on an adequate site investigation carried out prior to development. They do not apply to analysis of averaged, bulked or composited samples, nor to sites which have already been developed.
3 Many of these values are preliminary and will require regular updating. They should not be applied without reference to the current edition of the report 'Problems Arising from the Development of Gas Works and Similar Sites'
4 If all sample values are below the threshold concentrations then the site may be regarded as uncontaminated as far as the hazards from these contaminants are concerned, and development may proceed. Above these concentrations, remedial action may be needed, especially if the contamination is still continuing. Above the action concentration, remedial action will be required or the form of development changed.

required. A comprehensive list is given by Moen *et al.* (1986), shown in Table 3.3.

In Germany, the Federal Water Law is the driving force in the work of establishing some criteria for contaminated soils. In Hamburg, threshold values for toxic metals above which further investigation is necessary before site redevelopment takes place are being used, and a similar approach is being made in Baden-Württemburg. Other work is aimed at establishing thresholds for soil concentrations of contaminants which avoid damage to plants or result in unacceptably high accumulations of toxic metals in edible plants (Finnecy, 1987).

At present there is no consensus between the different member countries of the EEC about which figures should be adopted. A further problem is the lack of any significant toxicological data with which

to support the implementation of the threshold figures. Although the present figures are better than guess-work, there is still much research to be done before agreement can be reached and they can be applied with confidence.

Under the provisions of the Resource Conservation and Recovery Act, (RCRA) 1976, a waste material is defined as toxic in the USA if any contaminant in the 1:20 aqueous extract exceeds 100 times the National Interim Drinking Water Standard. In California, toxic waste is further defined as hazardous if it has:

(a) an acute oral LD50 of less than 5000 mg kg^{-1};
(b) an acute dermal LD50 of less than 4300 mg kg^{-1};
(c) an acute 8-h inhalation LC50 of less than 10 000 ppm;

Table 3.3 Guidelines for soil and groundwater contamination in the Netherlands (Moen *et al.*, 1986)

Substance	Concentration in					
	Soil (mg kg^{-1} dry weight)			Groundwater ($\mu g \, l^{-1}$)		
	A	B	C	A	B	C
Inorganic pollutants						
NH (as N)	–	–	–	200	1 000	3 000
F (total)	200	400	2 000	300	1 200	4 000
CN (total free)	1	10	100	5	30	100
CN (total complex)	5	50	500	10	50	200
S (total)	2	20	200	10	100	300
Br (total)	20	50	300	100	500	2 000
PO (as P)	–	–	–	50	200	700
Aromatic compounds						
Benzene	0.01	0.5	5	0.2	1	5
Ethyl benzene	0.05	5	50	0.5	20	60
Toluene	0.05	3	30	0.5	15	50
Xylene	0.05	5	50	0.5	20	60
Phenols	0.02	1	10	0.5	15	50
Aromatics (total)	0.1	7	70	1	30	100
Polycyclic aromatic compounds (PCAs)						
Napthalene	0.1	5	50	0.2	7	30
Anthracene	0.1	10	100	0.1	2	10
Phenanthrene	0.1	10	100	0.1	2	10
Fluoranthene	0.1	10	100	0.02	1	5
Pyrene	0.1	10	100	0.02	1	5
Benzo(a)pyrene	0.05	1	10	0.01	0.2	1
Total PCAs	1	20	200	0.2	10	40

Table 3.3 (*continued*)

	Soil (mg kg^{-1} dry weight)			Groundwater (µg l^{-1})		
	Concentration in					
Substance	A	B	C	A	B	C
Chlorinated organic compounds						
Aliphatic chlor. comp. (indiv.)	0.1	5	50	1	10	50
Aliphatic chlor. comp. (total)	0.1	7	70	1	15	70
Chlorobenzenes (indiv.)	0.05	1	10	0.02	0.5	2
Chlorobenzenes (total)	0.05	2	20	0.02	1	5
Chlorophenols (indiv.)	0.01	0.5	5	0.01	0.3	1.5
Chlorophenols (total)	0.01	1	10	0.01	0.5	2
Chlorinated PCA (total)	0.05	1	10	0.01	0.2	1
PCB (total)	0.05	1	10	0.01	0.2	1
EOCl (total)	0.1	8	80	1	15	70
Pesticides						
Organic chlorinated (indiv.)	0.1	0.5	5	0.05	0.2	1
Organic chlorinated (total)	0.1	1	10	0.1	0.5	2
Pesticides (total)	0.1	2	20	0.1	1	5
Other pollutants						
Tetrahydrofuran	0.1	4	40	0.5	20	60
Pyridine	0.1	2	20	0.5	10	30
Tetrahydrothiophene	0.1	5	50	0.5	20	60
Cyclohexanone	0.1	6	60	0.5	15	50
Styrene	0.1	5	50	0.5	20	60
Fuel	20	100	800	10	40	150
Mineral oil	100	1 000	5 000	20	200	600

(d) an acute 96-h LC50 of less than 500 mg l^{-1} measured in soft water with specified conditions and species;

(e) a content of 0.001% by weight (10 ppm) of any of 16 specified known carcinogenic organic chemicals:

(f) poses a hazard to human health or the environment because of its carcinogenicity, acute toxicity, chronic toxicity, bioaccumulative properties or persistence in the environment;

(g) a content of extractable or total bioaccumulative toxic substances which exceed soluble or total threshold limits;

(h) a content of one or more materials with a 8-h LC50 or LCLo of less than 10 000 ppm and the LC50 and LCLo is exceeded in the head space vapour;

(i) or it is a listed hazardous waste, designated as toxic.

LD50 and LC50 are the lethal dose and lethal concentration respectively which kill 50% of the population of laboratory test animals and LCLo is the lowest published lethal concentration.

Threshold limits and maximum acceptable limits have been designated for 40 persistent and bioaccumulative substances. Some substances are defined as extremely hazardous wastes; these include cyanides, hydrogen sulphide and parathion, with correspondingly lower threshold limits than those listed in the previous paragraph (Kingsbury & Ray, 1986).

It will readily be appreciated that the presence of waste material in soils of urban areas raises a number of questions about how to deal with them. Inert, non-toxic wastes pose little difficulty but where wastes are toxic, problems emerge both of a financial and practical nature. The first problem is that virtually all industrial sites have been polluted to some degree, but the extent and patterns of contamination are complex, time-consuming and costly to ascertain. Secondly, the clean-up of all sites, as is the declared aim in the Netherlands, requires an all-embracing survey of potential sites including those in current use. Thirdly, there is a problem of the order of priority between sites for reclamation. A further difficulty is that the number of trigger thresholds established is few and the list of potentially dangerous soil contaminants is very large. Even those established standards often lack the backing of toxicological knowledge, being based mainly on the availability to plants and the effects of passing toxic elements along the food chain (Finnecy, 1987). It is extremely important that further research is done to enable a better understanding of the problem of the relationships of waste materials in soils to the human population and ecosystems generally.

Conclusions

Urban soils can be found which have natural profiles, similar to soils with the same conditions of formation in rural areas. However, it is more likely that they will be modified by the addition of materials not normally considered to be soil and in some cases may be composed almost entirely of waste materials. The pore spaces of urban soils may contain liquids or gases derived from the decomposition of waste materials.

The influence of mankind upon soils has been profound and has steadily increased in intensity throughout historical time. Nowhere has this influence been more strongly felt than in the urban areas. Bidwell & Hole (1965) present a strong case for the consideration of mankind as a factor in soil formation and indicate under the headings of Jenny's (1941) five soil-forming factors, the beneficial and detrimental effects of our actions. Most of their comments apply to the manipulation of soils for agricultural production with only one or two points made about urban soils. Yaalon & Yaron (1966) coined the term meta-pedogenesis to cover those changes imposed upon the natural soil profile by man. Clearly, some of the changes envisaged by

these authors have taken place in urban soils, but additionally many waste materials have been added, either by accident or design, which are of significance when considering their characteristics and the use to which urban soils are put.

The effect of the presence of wastes in soils is largely unknown; some may be considered to be detrimental, such as the addition of toxic metals and other hazardous materials. Other wastes may be considered advantageous, such as the addition of organic wastes to improve the organic-matter content or basic slags to boost growth on grasslands. Furthermore, amenity soils have been created from some waste materials where no soil existed previously. Such soils occur in many of our cities and the plants they support beautify the urban landscape and make it a more pleasant place in which to live.

References

Barber, E.G. (1975). *Win Back the Acres: the Treatment and Cultivation of pfa Surfaces*. Central Electricity Generating Board, London.

Barnes, G.E. (1986). The effects of groundwater flow pattern on the concentration of soluble sulphates. In Assink, J.W. & van den Brink, W.J. (eds), *Contaminated Soil*, pp. 115–121. Martinus Nijhoff, Dordrecht.

Beckett, M.J. & Simms, D.L. (1986). Assessing contaminated land: UK policy and practice. In Assink, J.W. & van den Brink, W.J. (eds), *Contaminated Soil*, pp. 285–293. Martinus Nijhoff, Dordrecht.

Bidwell, O.W. & Hole, F.D. (1965). Man as a factor in soil formation. *Soil Science* **99**, 65–72.

Bins-Hoefnagels, I.J.M. & Molenkamp, G.C. (1986). Case studies of soil pollution in some Dutch urban areas. In Assink, J.W. & van den Brink, W.J. (eds), *Contaminated Soil*, pp. 769–779. Martinus Nijhoff, Dordrecht.

Bradley, R.G.V. & Rimmer, D.L. (1988). Dredged materials – problems associated with their use on land. *Journal of Soil Science* **39**, 469–482.

Bradshaw, A.D. & Chadwick. M.J. (1980). *The Restoration of Land*. Studies in Ecology Vol. 6. Blackwell Scientific Publications, Oxford.

Bridges, E.M. (1969). Eroded soils in the Lower Swansea valley. *Journal of Soil Science* **20**, 236–245.

Bridges, E.M. (1984). Desecration and Restoration in the Lower Swansea Valley. In *Management of Uncontrolled Hazardous Waste Sites*, pp. 553–559. Hazardous Materials Control Research Institute, Silver Spring, Maryland.

Bridges, E.M. (1987). *Surveying Derelict Land*. Clarendon Press, Oxford Science Publications, Oxford.

Bridges, E.M. (1988). *Healing the Scars: Derelict Land in*

Wales, University College of Swansea, Gomer Press.

Bridges, E.M., Evans, A.T. & Leech, D.J. (1986). *An Investigation into the Availability and Needs for Cover Material on Landfill Sites with Special Reference to the South West Wales Area*. WEP/126/100/2 Welsh Office.

Building Research Establishment (1981). Concrete in sulphate bearing soils and groundwater. *BRE Digest* No. 250. DoE, London.

Carpenter, R.J. (1986). Redevelopment of land contaminated by methane gas: the problems and some remedial techniques. In Assink, J.W. & van den Brink, W.J. (eds), *Contaminated Soil*, pp. 747–759. Martinus Nijhoff, Dordrecht.

Chumbley, C.G. (1971). *Permissable Levels of Toxic Metals in Sewage Used on Agricultural Land*. ADAS Advisory Paper No. 10. MAFF.

CONCAWE (1983). *A Field Guide to Inland Oil Spill Clean-up Techniques*. Report No. 10/83. Oil Companies European Environment and Health Council. Roepers Drukkerig, The Hague.

Cope, C.B., Fuller, W.H. & Willets, S.L. (1983). *The Scientific Management of Hazardous Wastes*. Cambridge University Press, Cambridge.

Dahl, T.O. (1986). Occidental Chemical Company at Lathrop, California, a groundwater/soil contamination problem and a solution. In Assink, J.W. & van den Brink, W.J. (eds), *Soil Contamination*, pp. 793–807. Martinus Nijhoff, Dordrecht.

Davies, B.E. (1971). Trace element content of soils affected by base metal mining in the west of England. *Oikos* **22**, 366–372.

Davies B.E. & Roberts, L.J. (1978). The distribution of heavy metal contaminated soils in northeast Clwyd, Wales. *Water, Air and Soil Pollution* **9**, 507–518.

Davies, B.E., Elwood, P.C., Gallacher, J. & Ginnever, R.C. (1985). The relationship between heavy metals in garden soils and house dusts in an old lead mining area of North Wales, Great Britain. *Environmental Pollution Series B* **9**, 255–266.

Davies, R.L. (1969). Environmental monitoring and control. In Bromley, R.D.F. & Humphrys, G. (eds), *Dealing with Derelict Land*, pp. 73–87. University College of Swansea.

Department of the Environment (1977). *Solvent Waste (Excluding Halogenated Hydrocarbons)*. Waste Management Paper No. 14, HMSO, London.

Department of the Environment (1981). *Special Wastes: a Technical Memorandum Providing Guidance on their Definition*. Waste Management Paper No. 23, HMSO, London.

Finnecy, E.E. (1987). Impacts on soils related to industrial activities. 2. Incidental and accidental soil pollution. In Barth, H. & L'Hermité, P. (eds), *Soil Protection*, pp.

259–280. Elsevier, The Netherlands.

Gemmell, R.P. (1973). Revegetation of land polluted by a chromate smelter. 1. Chemical factors causing substrate toxicity in chromate smelter waste. *Environmental Pollution* **5**, 181–197.

Goodman, G.T. & Roberts, T.M. (1971). Plants and soils as indicators of metals in the air. *Nature, London* **231**, 287–292.

Goodman, G.T. & Smith, S. (1975). Relative burdens of airborne metals in South Wales. In *Report of a Collaborative Study on Certain Elements in Air, Soil, Plants, Animals and Humans in the Swansea–Neath–Port Talbot Area Together with a Report on a Moss-bag Study of Atmospheric Pollution across South Wales*. Welsh Office, Cardiff.

Gutt, W., Nixon, P.J., Smith, M.A., Harrison, W.H. & Russell, A.D. (1974). *A Survey of the Locations, Disposal and Prospective Uses of the Major Industrial By-products and Waste Materials*. CP 19/74 Building Research Establishment, DOE, London.

Heeps, K.D. & Pike, E.R. (1980). Reclamation of a disused sewage farm. In *Reclamation of Contaminated Land*. Society of Chemical Industry, London.

Highfield, R. (1985). Common clay detoxifies dioxin. *New Scientist* 12 September.

Hodgson, D.R. & Townsend, W.N. (1973). The amelioration and revegetation of pulverized fuel ash. In Hutnik, R.J. & Davis, G. (eds), *Ecology and Reclamation of Devastated Land*, pp. 247–271. Gordon and Breach, New York.

Interdepartmental Committee on the Redevelopment of Contaminated Land, (1979). *Redevelopment of Gasworks Sites*. Interdepartmental Committee on the Redevelopment of Contaminated Land. ICRCL 18/79 DoE, London.

Interdepartmental Committee on the Redevelopment of Contaminated Land, (1980). *Notes on the Redevelopment of Scrapyards and Similar Sites*. Interdepartmental Committee on the Redevelopment of Contaminated Land. ICRCL 42/80. DoE. London.

Interdepartmental Committee on the Redevelopment of Contaminated Land, (1987). *Guidance on the Assessment and Redevelopment of Contaminated Land*. Interdepartmental Committee on the Redevelopment of Contaminated Land. ICRCL 59/83 (Second Edition) DoE, London.

James, S.C., Kinman, R.N. & Nutini, D.L. (1985). Toxic and Flammable Gases. In Smith, M.A. (ed.), *Contaminated Land*, pp. 207–255. NATO Challenges of Modern Society Volume 8. Plenum Press, London.

Jenny, H. (1941). *Factors of Soil Formation*. McGraw-Hill, New York.

Kingsbury, G.L. & Ray, R.M. (1986). *Reclamation and*

Redevelopment of Contaminated Land: 1. US Case Studies. EPA/600/2–86/066 Cincinnati.

Lowe, G.W. (1980). GLC development at Thamesmead; investigation and reclamation of contaminated land. In *Reclamation of Contaminated Land*. Society of Chemical Industry, London.

Lutz, H.J. & Chandler, R.F. (1947). *Forest Soils*. John Wiley, New York.

McCalla, T.M., Peterson, J.R. & Lue-Hing, C. (1977). Properties of agricultural and municipal wastes. In Elliott L.F. & Stevenson, F.J. (eds), *Soils for Management of Organic Wastes and Waste Waters*, Soil Science Society of America. Madison, Wisconsin.

Mackenzie, D. (1985). Dioxin: still looking for the bodies. *New Scientist* 26 September.

Moen, J.E.T., Cornet, J.P. & Evers, C.W.A. (1986). Soil protection and remedial actions: criteria for decision making and standardization of requirements. In Assink, J.W. & van den Brink, J.W. (eds), *Contaminated Soil*, pp. 441–448. Martinus Nijhoff, Dordrecht.

Paigen, B., Goldman, L.R., Highland, J.H., Magnant, M.M. & Steegman, A.T. (1985). Prevalence of health problems in children living near Love Canal. *Hazardous Waste and Hazardous Materials* 2, 23–43.

Porteous, A. (1975). Domestic refuse: its composition, properties, recovery potential and disposal methods, present and future. In Benn, F.R. & McAuliffe, C.A. (eds), *Chemistry and Pollution* pp. 24–47. Macmillan, London.

Porteous, C, (n.d.), *Pioneers of Fertility*. Clareville Press, London.

Richardson, J.A. (1975). Physical problems of growing plants on colliery waste. In Chadwick, M.J. & Goodman, G.T. (eds), *The Ecology of Resource Degradation and Renewal*, pp. 275–285. Blackwell Scientific Publications, Oxford.

Rimmer, G. (1976). Putting solid ground beneath Corby's feet. *Surveyor* **147**, 24–25.

Sanning, D.E. (1980). *Remedial Action Technologies for Uncontrolled Hazardous Waste Sites – Needs and Solutions*. Expert seminar on hazardous waste problem sites. OECD, Paris.

Sax, N.I. & Lewis, R.J. (1989). *Dangerous Properties of Industrial Materials*. (3 volumes) Van Nostrand Reinhold, New York.

Schuuring, C. (1981). Dutch Dumps. *Nature, London* **289**, 340.

Smith, M.A. (1982) *Register of Important Sites*. NATO/CCMS Pilot Study of Contaminated Land. Building Research Establishment. Garston, Beds.

Steif, K. (1984). Remedial action for groundwater protection; case studies within the Federal Republic of Germany. In *Management of Uncontrolled Hazardous Waste Sites*, pp. 565–568. Hazardous Materials Control Research Institute, Silver Spring, Maryland.

Sumner Report (1978). *Co-operation Programme of Research on the Behaviour of Hazardous Wastes in Landfill Sites*. HMSO, London.

Thornton, I. (1986). Metal contamination of soils in UK urban gardens: implications to health. In Assink, J.W. & van den Brink, J.W. (eds), *Contaminated Soil*, pp. 203–209. Martinus Nijhoff, Dordrecht.

Townsend, W.N. & Gilham, E.W.F. (1975). Pulverized fuel ash as a medium for plant growth. In Chadwick, M.J. & Goodman, G.T. (eds), *The Ecology of Resource Degradation and Renewal*, pp. 287–304. Blackwell Scientific Publications, Oxford.

Trevelyan, G.M. (1947). *English Social History*. Longmans, Green and Co., London.

Webb, J.S., Thornton, I., Howarth, R. J. & Lowenstein, P.L. (1978). *The Wolfson Geochemical Atlas of England and Wales*. Oxford University Press, Oxford.

Welsh Office (1983). *The Halkyn Mountain Survey*. Welsh Office, Cardiff.

Yaalon D.H. & Yaron, B. (1966). Framework for man-made soil changes – an outline of meta-pedogenesis. *Soil Science* **102**, 272–277.

4 Metal contamination of soils in urban areas

I. THORNTON

Introduction

Until very recently soil scientists have been primarily concerned with soil as a basis for agriculture and food production. Classification systems and research into the physical and chemical properties of soils have been focused towards the requirements of farming and forestry and to a degree towards the understanding of natural ecosystems; the urban environment, in which the majority of the population lives and comes into contact with the soil, has been almost totally neglected.

A question addressed in the majority of the chapters concerns the degree to which our accumulated knowledge of relatively undisturbed rural soils can be applied to the disturbed urban and suburban environment.

Barrett (1987) in a recent review of research in urban ecology concludes that 'urban areas differ from rural ones in both the scale and intensity of human impacts . . . ' and states that 'the continuing cycle of construction, use and renewal of urban structures leads to far higher rates of change than is common in non-urban environments. The physical (and chemical) environment of cities is profoundly affected by almost every kind of human activity, from deliberate acts of construction, management or vandalism to accidental or incidental pollution'. A detailed appraisal of the characteristics of urban soils by Craul (1985) points out that soils in urban and suburban areas are frequently disturbed and subjected to mixing, filling and contamination with heavy metals, herbicides and pesticide residues.

The history of land use in urban areas is often difficult to ascertain. Records of previous industrial use or waste disposal are frequently poor or do not exist. Major industrial sources of pollution from the last century may be masked by the presence of post-war housing; sequences of changing land-use and transport of fill materials may be reflected in the presence of contaminated materials below the surface. Urban development coupled with the presence of industrial activities within urban areas leads to varying degrees of soil contamination with one or more materials.

An attempt to produce an inventory of worldwide emissions from industrial and domestic sources suggests that soils are receiving large quantities of trace metals from a variety of industrial wastes, the disposal of ash residues from coal combustion and the general wastage of commerical products on land (Nriagu & Pacyna, 1988). The authors calculate that if the total metal inputs were dispersed uniformly, annual rates of deposition would vary from $1\,g\,ha^{-1}$ for Cd and Sb to about $50\,g\,ha^{-1}$ for Pb, Cu and Cr and over $65\,g\,ha^{-1}$ for Zn and Mn. Naturally deposition is not uniform and industrial countries such as Britain will receive larger inputs.

Changes in the chemical nature of the soil brought about by the addition of pollutants may present a hazard to construction works, may lead to adverse interactions with building materials, toxicity to soil flora and fauna including ornamental plants and trees and garden crops, or may lead to the accumulation of toxic substances in vegetables making them unsuitable for human consumption. Flux between surface soils and dusts may influence exposure of the population and particularly young children to toxic substances and present a hazard to human health.

The complex nature of urban land and its many uses present problems in assessing the extent and degree of contamination. Sampling strategies for the assessment of the suitability of soil for future development usually have to be site specific and will depend on the intended land use.

We are reminded by Barrett (1987) that urban areas are characterized 'by the presence of large numbers of buildings, roads . . . which form an impermeable, largely sterile, covering over much of the land surface . . . . In general central areas will have at least 80% cover, while suburban areas will have around 50% cover'. Exposed land surfaces include:

(a) grass verges alongside roads,
(b) private gardens and allotments,
(c) public parks and gardens,

(d) playing fields and golf courses,

(e) cemeteries,

(f) demolition and building sites,

(g) wasteland and derelict land,

(h) rubbish tips and spoil heaps,

(i) railway land,

(j) canal and river banks including disused docklands,

(k) woodland, heath and common land,

(l) farmland enclosed by urban developments.

Each of these units is subjected to varying degrees of contamination depending on the location, the past and present land use and the proximity to pollution sources.

The need for a well-defined programme for the protection of soils was fully discussed at a symposium organized by the Commission of the European Communities (Barth & L'Hermite, 1987). Major impacts arising from changes in land use over the past 200 years include the mass sterilization of land resources for industrialization and urbanization, and the growth in airborne pollution and the effects of fallout on vegetation cover, soil quality and the hydrological systems (Moss, 1987).

Sources of contamination

Classification systems for derelict land, with the implication that much of this will fall within the urban environment, have been reviewed by Hollis (this volume) and previously by Bridges (1987). These take into account the composition of the surface (e.g. burnt shale) (Beaver, 1946), present day and past activities (e.g. gasworks) (Collins & Bush, 1969), and causes of despoilation (e.g. power stations) (Downing, 1977). More recently Haines (1987) has attempted to classify categories of contamination according to the site use, on the assumption that site use, pollutants and land contamination are linked. The classes proposed are:

(a) definitely contaminated;

(b) probably contaminated;

(c) potentially contaminated;

(d) pre-1931 housing where lead paint might be expected;

(e) mixed land use with many small, potentially contaminated, sites;

(f) definitely uncontaminated;

(g) unknown;

(h) previously contaminated sites that are now in 'sensitive' use.

These categories are perhaps of a subjective nature though are aided by the publication of guidelines and 'trigger concentrations' for a range of toxic substances issued by the Interdepartmental Committee on the Redevelopment of Contaminated Land (ICRCL, 1983, 1987). The need for such guidelines will be addressed in a later section of this chapter. However, Bridges (1987) emphasizes that 'classification according to the degree of contamination is not easy as there is a wide range of possible contaminants of different degrees of toxicity'.

The main sources of metal contamination in urban areas may conveniently be characterized according to several broad descriptions of past and present day activities. However, before listing these, it is perhaps helpful to mention certain general factors which influence the extent and degree of such contamination. First, individual pollutants may be dispersed over either a wide area from diffuse sources, such as vehicle emissions and ammonia from farm livestock, or a localized area from point sources such as a metal works. Second, dispersion may be in the form of atmospheric gases and particulate materials, liquid effluents from factories and sewage works and *in situ* site contamination from operational activities and from disposal of waste products. Finally, dispersion may either result from controlled emissions such as those from a smelter stack or effluent pipe which are required to meet standards applied by HM Pollution Inspectorate or the local water authority or from uncontrolled, accidental or fugitive emissions and spillages which are difficult to quantify.

HOUSEHOLD ACTIVITIES

Accumulation of heavy metals in household gardens is well documented (Davies, 1978; Thornton *et al.*, 1985), lead concentrations in surface (0–15 cm) soils increasing with the age of the house. Sources of metals in garden soils include the disposal of fossil fuel residues (ash and soot) and of household refuse, bonfires, and fragments of lead-containing paints, long-term application of phosphatic fertilizers (Cd) and deposition of atmosphere particulates from industrial processes and vehicle emissions.

WASTE DISPOSAL

Problems associated with waste disposal and management are reviewed in depth in the 11th Report of the Royal Commission on Environmental Pollution (1985). Toxic metals are present in domestic refuse in varying concentrations and frequently mixing of domestic and industrial wastes results in isolated pockets of these and other hazardous materials including phenols, cyanides and asbestos.

Some 90% of refuse is disposed of in landfill sites, some in urban areas and some destined for future development for housing, etc. These is usually little loss of metallic components from landfill sites into neighbouring soils and watercourses, though leachates may contain other more soluble toxic constituents, be extremely acidic and contaminate ground water.

The other 10% of refuse is processed in municipal incinerators, from which it has been estimated that as much as 6 t of cadmium and 115 t of lead per year are discharged into the UK atmosphere (Wadge and Hutton, 1987). However, in a study around an incinerator at Edmonton, North London, which processes 4×10^5 t of waste per year, there was little evidence of extensive Cd and Pb contamination downwind, though surface soils within 0.2 km of the stack were enriched in Cd (12-fold) and Pb (2-fold) compared with nearby areas (Hutton *et al.*, 1988).

Land previously occupied by sewage works may be redeveloped and may include old filter beds, lagoons, etc. and extensive areas in which effluent had been dispersed. Such activities can lead to soils being contaminated with metal wastes, combustible materials and pathogens. Bridges (1987) refers to detailed studies undertaken at the Beaumont Leys Estate, Leicester, a sewage farm from 1890 to 1964, required for urban development. An area in which sludge had been spread exceeded the 'trigger' concentration for zinc equivalent (ICRCL, 1983) and, as a result, was designated as an area of public open space; less contaminated areas were developed for housing and industrial premises.

TRANSPORT

Roads and motor vehicles

Nearly all the lead in the air in Britain comes from the exhaust gases of petrol engines (Royal Commission on Environmental Pollution, 1983). In 1981, some 9.7 thousand tonnes of lead were used in petrol of which about 75% was released to the atmosphere (Chamberlain *et al.*, 1978). The rate of deposition decreased rapidly with distance from the road (Fig. 4.1). It has been estimated that about 10% of emitted lead is deposited within 100 m of a road and that the remainder can be transported over considerable distances (Little & Wiffen, 1978).

Page & Ganje (1970) attempted to relate accumulation of lead in Californian surface soils over a period of 35–50 years in regions of high and low motor vehicle traffic density and estimated an accumulation of 15–36 $\mu g\,g^{-1}$ over 40 years in the

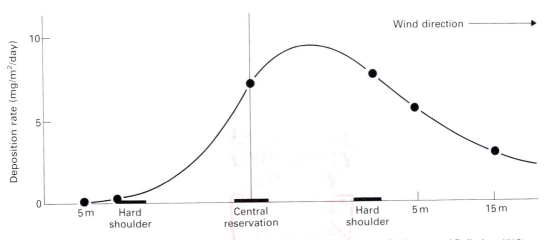

Fig. 4.1 Deposition of lead on grass beside a motorway (from Royal Commission on Environmental Pollution, 1983).

Table 4.1 Lead contents of soil at three distances from typical main roads, calculated from data in Smith (1976)

Distance from road	Geometric mean	95% probability range	Number of samples
(m)		(μg Pb g^{-1} soil)	
< 10	192	18–2017	20
15	161	50– 511	6
> 30	53	14– 203	17

high traffic area and no measurable amount in the low area. It is now widely accepted that the settling out of lead-rich aerosols derived from exhaust fumes of cars results in an increased concentration of lead in surface soil and this is demonstrated by the data in Table 4.1 calculated from Smith (1976). This topic is discussed in detail in a review by Davies & Thornton (1990). Although the lead content of petrol was reduced from 0.4 to 0.15 g l^{-1} in 1986, resulting in levels of lead in air falling by approximately 50% and no doubt a reduction in the deposition of airborne lead to soil, it must be stressed that lead in soil is virtually immobile, that contamination is in essence a permanent phenomenon, and that concentrations of lead resulting from many years deposition will remain.

METALLIFEROUS MINING AND SMELTING

In Britain, the oldest and most extensive sources of metal pollution are related to the history of metalliferous mining which commenced in Roman and earlier times, thrived until the end of the nineteenth century, and then declined rapidly. Peak output was in the mid-nineteenth century, when Britain produced 75% of the world's copper, 60% of the tin, and 50% of the lead. It has been estimated that in excess of 4000 km² of agricultural land (and many urban areas) in England and Wales are contaminated with one or more metals due to this legacy from our forebears (Thornton, 1980). This subject has been reviewed in detail by Thornton & Abrahams (1984). Village and town communities developed around the mines and smelters and thus have often expanded in area onto reclaimed land previously despoiled with mine waste and smelter slag. Mine shafts may even occur within the built-up area as in Shipham, Somerset (Thornton *et al.*, 1980). Metal contamination of soils may be considerable, with as much as 1% or more lead recorded in surface garden soils in the village of Winster, Derbyshire (Barltrop *et al.*, 1975). In Cornwall for example, more than 600 mines have operated in the past, mainly within a belt 75 miles long and 10 miles wide (Fig. 4.2). Within this area soils are contaminated to varying degrees with tin, copper and arsenic (Thornton *et al.*, 1986). A survey of arsenic in gardens in the towns of Redruth and Hayle showed arsenic concentrations in surface soils to range widely up to nearly 900 ppm As (Xu & Thornton, 1985), compared to a median value of about 10 ppm for uncontaminated agricultural soils (Archer & Hodgson, 1987). Other areas affected are the Mendips (Pb, Zn, Cd), north and central Wales (Pb, Zn, Cd), and the north and south Pennines (Pb, Zn, Cd, F).

Primary metal smelters at Avonmouth and Capper Pass, have resulted in some contamination of surface soils through stack and fugitive emissions, transport of metal concentrates and accidental wheel drag out. Similarly secondary metal smelters will always give rise to a limited amount of pollution, leading in some cases to public concern.

MANUFACTURING INDUSTRIES

These are many and of various kinds. Those with which metal contamination is associated include:
(a) engineering, vehicle construction;
(b) printing;
(c) paint works (lead, chromium, antimony, etc.);
(d) leather industry (chromium).

SCRAP YARDS

Storage, processing, recycling and disposal of scrap materials will always result in site contamination and often soils will be contaminated with a range of toxic substances. Heavy metals and metalloids, waste oil, and other organic and inorganic compounds may be present and removal of contaminated soil is usually necessary prior to redevelopment.

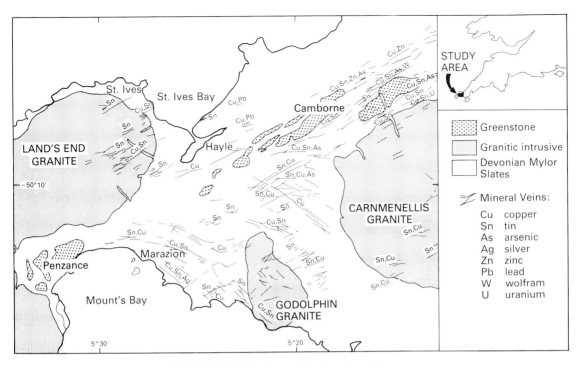

Fig. 4.2 Geology and mineralogy within the Hayle–Cambourne district, Cornwall.

Sources of trace elements and metals in soils

The main sources of trace elements in soils are the parent materials from which they are derived. Usually this is weathered bedrock or overburden that has been transported by wind, water or glacial activity. Overburden may be local or exotic, though in Britain, transported material is mainly of local origin. Ninety-five percent of the earth's crust is made up of igneous rocks and 5% sedimentary rocks; of the latter about 80% are shales, 15% sandstones and 5% limestones (Mitchell, 1964). Sedimentary rocks tend to overlie the igneous rocks from which they were derived and hence are more frequent in the surface weathering environment. The degree to which trace elements in igneous rocks become available on weathering depends on the type of minerals present. The more biologically important trace elements such as Cu, Co, Mn and Zn occur mainly in the more easily weathered materials such as hornblende and olivine (Mitchell, 1974). Of the sedimentary rocks, sandstones contain minerals that weather slowly and usually contain only small amounts of trace elements. On the other hand, shales may be of organic or inorganic origin, and usually contain large amounts of trace elements (Mitchell, 1964). Black shales in particular are enriched in a number of elements including Cu, Pb, Zn, Mo and Hg, sometimes at levels deleterious to plant and/or animal health.

Soils derived from these parent materials often tend to reflect the chemical compositions of the parent material. This is illustrated by the data for Scottish soils shown in Table 4.2 (Berrow & Ure, 1986). Soils developed from the weathering of coarse-grained sediments such as sands and sandstones, and from acid igneous rocks such as rhyolites and granites tend to contain smaller amounts of nutritionally-essential elements such as Cu and Co, than those derived from fine-grained sedimentary rocks such as clays and shales and from basic igneous rocks. Potentially toxic amounts of trace elements in soils may be derived as the result of weathering of metal-rich source rocks (Table 4.3). For example, some calcareous soils developed from interbedded shales and limestones of the Lower Lias formation (Juras-

Table 4.2 Arithmetic mean contents ($\mu g\,g^{-1}$) in B-horizon samples of Scottish soil profiles from different associations (from Berrow & Ure, 1986)

Association	Parent material	% of land area	No. samples	Co	Cu	As	Ni
Igneous and metamorphic materials							
Leslie	Ultra basic rocks	0.12	13	88	27	2.3	1 540
Incsh	Basic igneous rocks	0.67	21	35	29	1.1	57
Darleith	Basaltic rocks	3.53	58	30	25	1.3	51
Sourhope	Intermediate lavas	1.71	42	16	22	1.4	41
Foudland	Slates	3.25	67	14	21	1.4	36
Strichen	Dalradian schists	7.98	102	10	14	0.29	29
Tarves	Mixed igneous and metamorphic rocks	2.07	60	25	33	3.9	57
Countesswells	Granite rocks	5.75	59	5.5	7.7	2.7	22
Corby/Boyndie	Fluvioglacial sands and gravels	3.08	81	5.0	7.5	0.60	19
Sedimentary materials							
Stirling	Silts and clays	0.54	19	17	16	1.1	47
Ettrick	Greywackes and shales	9.26	86	16	22	1.4	55
Rowanhill	Carboniferous drifts	3.06	66	15	21	1.6	46
Balrownie	Lower ORS sandstones	1.83	89	16	20	0.81	46
Thurso	Middle ORS flagstones	1.35	36	8.3	17	1.9	35
Hobkirk	Upper ORS drifts	0.75	16	7.6	8.0	1.0	28
Canisbay	Middle ORS sandstones	0.39	37	6.6	12	1.7	30
Cromarty	ORS drifts	0.31	58	2.9	3.9	0.8	22
Organic Soils							
Peat	Peat profiles	9.94	351	2.1	6.5	2.3	8.0

sic) in southwest England contain 20 ppm Mo or more, and are associated with molybdenosis and molybdenum-induced copper deficiency in grazing cattle.

The influence of parent materials on the total content and form of trace elements in soils is modified to varying degrees by pedogenetic processes that may lead to the mobilization and redistribution of trace elements both within the soil profile and between neighbouring soils. In the United Kingdom and similar temperate areas, most of the soils are relatively young and the parent material remains the dominant factor. Under tropical climates and on more mature land surfaces, such as those in Australia, weathering processes have been vigorous or of much greater duration and relationships between the chemical composition of the original parent materials and the soil may be completely overridden by the mobilization and secondary distribution of chemical elements and the formation of secondary minerals.

The processes of gleying, leaching, surface organic matter accumulation and podzolization, together with soil properties such as reaction (pH) and redox potential (Eh) may affect the distribution, the form and the mobility of trace elements in the soil. Trace elements, including Se, are often leached from the surface layers of podzols and enriched in the B horizon (Smith, 1983).

Soils in mineralized areas are often enriched in the ore metals and in Britain frequently contain high concentrations of one or more of the elements Cu, Pb, Zn, Cd and As.

It is of interest that the concentrations of lead in agricultural soils in the USA appear lower than those in Britain. Holmgren *et al.* (1983) report a median

NATIONAL SOIL INVENTORIES

Comprehensive baseline data for several heavy metals, trace elements and major nutrients have recently been obtained for soils in England, Wales and Scotland.

Some 5800 topsoil (0–15 cm) samples were taken by the Soil Survey of England and Wales (now renamed Soil Survey and Land Research Centre) at 5 km intervals on a square sampling grid throughout England and Wales as part of a natural inventory of soils. These soils were analysed by the Soils and Plant Nutrition Department at Rothamsted Experimental Station for P, K, Ca, Mg, Na, Fe, Al, Ti, Zn, Cu, Ni, Cd, Cr, Pb, Co, Mo, Mn, Ba and Sr by ICP emission spectrometry and colour maps produced (McGrath *et al.*, 1986). These data are summarized for topsoils from Wales and central, southern and southwest England in Table 4.5 in which geometric mean values (*n* = 2776) for lead of 48 ppm, cadmium 0.9 ppm, zinc 85 ppm and copper 18 ppm (McGrath 1986, 1987) are appreciably smaller than those to be expected in urban soils.

A similar survey of Welsh soils, commissioned by the Welsh Office, was based on 722 A-horizon and 662 B-horizon samples, also collected by the Soil Survey of England and Wales, which were analysed in the laboratories of the Geography Department, University College of Wales, Aberystwyth by flame and flameless atomic absorption spectrophotometry

after extraction with aqua regia (Davies, 1986; Davies & Paveley, 1990). The results in Table 4.6 show smaller median values in surface soils for lead, cadmium, zinc and copper than those listed above; the difference between mean and median values recorded reflects contamination of some samples from mining and other industrial activities.

Total contents of some 20 elements have been determined in horizons of some 1000 soil profiles in Scotland, representing the various soil types, using spectrographic and atomic absorption methods; data for extractable soil contents for some elements are also available (Berrow & Ure, 1986). An arithmetic mean value of 24 ppm lead has been quoted for 3944 samples from 896 soil profiles in Scotland, which in turn is smaller than the median values listed in Tables 4.5 and 4.6 for England and Wales (Reaves & Berrow, 1984). It has since been proposed that atmospheric sources contribute a major portion of lead in uncultivated upland topsoils in south and central Scotland (Berrow *et al.*, 1987).

Further comprehensive data for agricultural soils are provided by Archer & Hodgson (1987) based on analysis for total and extractable trace elements on samples taken between 1973 and 1980 from farms selected for annual Surveys of Fertilizer Practice. The data (0–15 cm surface soils) are illustrated for total concentrations of arsenic, chromium and mercury in Table 4.7 and total and extractable lead in Table 4.8.

Table 4.5 Summary statistics of the concentrations of six metals in the topsoils of England and Wales (*n* = 2776) (from McGrath, 1987)

	Zn	Cu	Ni	Cd	Cr	Pb
			(mg kg^{-1} air dry soil)			
(a) *Untransformed data*						
Mean	103	23	26	1.2	45	75
Median	87	18	24	0.9	43	42
Standard deviation	111	27	17	2.4	27	338
Skewness	11	10	3	36	6	41
Kurtosis	153	147	25	1621	98	1936
(b) *Log$_{10}$-transformed*						
Geometric mean	85	18	21	0.9	38	48
Geometric deviation	1.8	1.8	2.1	2.2	1.8	2.00
Skewness	0.08	0.40	−1.20	−0.25	−1.46	1.57
Kurtosis	4.00	2.97	2.75	0.98	4.73	5.79

		Minimum	Maximum	Mean	Median	Mean/Medium
Pb	A	1.3	3369	17	35	2.0
	B	0.16	2095	33	17	1.9
Zn	A	4.7	2119	78	63	1.2
	B	0.04	1451	68	59	1.2
Cu	A	0.13	214	16	12	1.3
	B	0.09	65	13	11	1.2
Cd	A	0.01	15	0.50	0.29	1.7
	B	0.01	12	0.36	0.12	3.0
Co	A	0.14	2844	13	7.5	1.7
	B	0.14	43	10	9.3	1.1
Ni	A	0.42	169	16	14	1.1
	B	0.63	79	20	19	1.1
Mn	A	3.0	(0.5%)	2493	601	4.1
	B	1.7	(0.1%)	739	547	1.4
Fe	A	38	(0.12%)	14 243	12 408	1.1
	B	12	(0.10%)	16 150	13 649	1.2
pH	A	2.0	7.8	4.5	4.4	1.0
	B	2.3	7.0	4.3	4.0	1.1
OM	A	0.29	99	12	5.5	2.2
	B	< 0.10	98	5.5	2.3	2.4

Table 4.6 Summary statistics for 722 A-horizon and 662 B-horizon soil samples from profiles in Wales. Metal values are from a hot HCl/HNO_3 extraction and are as $mg\,kg^{-1}$ dry soil, pH was determined after equilibration for 30 min in 0.05 M $CaCl_2$ and per cent organic content (OM) was determined gravimetrically after ignition at 430°C (from Davies, 1986)

	As (total) ($mg\,kg^{-1}$)	Cr (total) ($mg\,kg^{-1}$)	Hg (total) ($mg\,kg^{-1}$)
Log-derived mean	11.0	42.4	0.09
Median	10.4	54.0	0.09
Overall range	1.0–140	4.0–160	0.01–2.12
'Normal' range	2.3–53	9.9–121	0.02–0.40
No. of samples	222	192	305

Table 4.7 Arsenic, chromium and mercury: total ($mg\,kg^{-1}$) (from Archer & Hodgson, 1987)

Frequency distribution of values

Range ($mg\,kg^{-1}$)	As (total) No. of samples	Cr (total) No. of samples	Range ($mg\,kg^{-1}$)	Hg (total) No. of samples
< 1	4	0	< 0.1	173
⩾ 1–< 5	11	2	⩾ 0.1–< 0.2	93
5–10	88	7	0.2–0.5	93
10–20	87	21	0.5–1.0	7
20–40	18	32	2.12	1
40–60	7	46		
60–80	2	63		
80–100	1	16		
100–120	1	3		
⩾ 120	3	2		

Table 4.14 Lead in blood, environmental samples, hand-wipes, diet and water

Sample	Units	n	Geometric mean	Percentiles 5th	95th
Blood	μg 100 ml^{-1}	97	11.7	6	24
Air					
playroom	μg m^{-3}	607	0.27	0.08	0.88
bedroom	μg m^{-3}	599	0.26	0.09	0.81
external	μg m^{-3}	605	0.43	0.12	1.53
Dust	μg g^{-1}	94	424	138	2093
Soil	μg g^{-1}	87	313	92	1160
Dust 'loading'	μg m^{-2}	93	60	4	486
Hand-wipes	μg	704	5.7	1.9	15.1
Diet (food and beverages)	μg week^{-1}	96	161	82	389
Water	μg l^{-1}	96	19	5	100

Table 4.15 Intake and uptake of lead by young children in Birmingham

Intake route	Pathway	Lead uptake (μg day^{-1})	Total uptake (%)
Inhalation	Air	1.1	3
Ingestion	Diet	12.2	34
	Dust	22.5	63
Total		35.8	100

Table 4.16 Pb, Zn and Cd in garden/allotment soils (μg g^{-1}) and in radish and lettuce grown directly in the soils (μg g^{-1}, dry matter) (from Thornton & Jones, 1984)

Location	Pb Soil	Radish	Lettuce	Zn Soil	Radish	Lettuce	Cd Soil	Radish	Lettuce
London	4100	23.3	23.5	1562	267	144	2.8	0.7	0.8
	2180	11.9	16.4	2182	333	309	1.8	0.7	0.8
Newcastle	840	12.2	9.4	506	65	97	1.4	0.4	0.6
upon Tyne	1608	7.3	8.5	660	52	101	1.8	0.5	0.6
Leeds	690	3.3	3.8	554	47	87	1.2	0.3	0.4
	136	3.2	4.1	190	26	76	0.6	0.6	0.7
Stoke on	224	1.6	4.4	276	127	166	0.8	0.3	2.3
Trent	90	1.5	1.7	174	50	99	0.6	0.2	0.8
Scunthorpe	108	1.4	2.0	258	42	81	< 0.1	0.2	0.4
	64	2.0	2.2	320	42	83	0.2	0.3	0.3
Control soil	92	–	–	160	–	–	0.5	–	–

lead in washed lettuce at the London and Newcastle sites was apparently associated with foliar uptake or soil splash, as concentrations of lead in plants grown in a control soil at these locations were large. Chamberlain (1983) has reviewed the subject of lead fallout onto crops and concludes that foliar deposition of airborne lead accounts for most of the lead in grasses and in other plants having a high leaf surface per unit mass. These criteria would apply to lettuce and other large-leaved garden crops such as cabbage, kale and spinach. In the present study up to 75% of lead, nickel and iron was removed from lettuce by washing. However, in spite of the elevated levels of metal in some of these soils, only lettuce and radish from London exceeded the statutory limits of 1 ppm Pb fresh weight. Lead derived from this source may be important when lead exposure from other sources is also above normal.

Another study was centred on an area of west Cornwall in which arsenic (together with tin and copper) had been mined. In July 1984, soils, vegetable and salad crops were sampled in 32 gardens in the towns of Hayle and Camborne as described by Xu & Thornton (1985). Concentrations of arsenic in topsoils (0–15 cm), lettuce, onion, beetroot, carrot, pea and bean are listed in Table 4.17. Total concentrations of arsenic in soils ranged widely (144–892 $\mu g\,g^{-1}$; geometric mean 322 $\mu g\,g^{-1}$), compared with normal soils in the UK (5–100 $\mu g\,g^{-1}$). Arsenic concentrations in the edible tissues of the six garden crops were not high, but were species-dependent with maximum concentrations in lettuce (geometric mean 0.85 $\mu g\,g^{-1}$ dry matter). Relationships between arsenic in soil and in beetroot, lettuce, onion and pea are significant. The authors derived regression equations based on soil arsenic to predict

arsenic concentrations in the vegetables and used ridge regression analysis to test the effect of other soil variables. Total soil iron reduced the uptake of arsenic by lettuce, and phosphorus increased arsenic uptake.

The statutory limit for arsenic in most foods offered for sale in the UK is 1 ppm fresh weight. In the study reported, although arsenic levels in vegetables were above those recorded elsewhere in the UK (MAFF, 1982), even in this 'geochemical hotspot' situation all the vegetables examined were below this permitted level and most were below 0.2 ppm as fresh weight.

CADMIUM EXPOSURE IN SHIPHAM, SOMERSET

The geochemical survey of England and Wales referred to earlier (Webb *et al.*, 1978) drew attention to several areas in which soils contain high concentrations of cadmium (see Fig. 4.5), usually associated with large amounts of zinc and sometimes lead. In the vicinity of Shipham, surface agricultural soils reclaimed from and near old mine workings ranged from 30 to 800 ppm Cd, compared with a normal value of less than 2 ppm Cd. Zinc was mined in and around this village from around 1700 to 1850, mainly as the ore calamine (smithsonite, $ZnCO_3$), which was used with copper to produce brass. Mineral veins run under parts of the village and several housing developments lie on land reclaimed from old workings. Results of a survey of metals in garden soils in Shipham and in a nearby control village, North Petherton, have been published by Thornton *et al.* (1980) and Moorcroft *et al.* (1982) and further discussed by Thornton (1988). Surface

Table 4.17 Concentrations of As in garden soils and vegetables ($\mu g\,g^{-1}$, dry matter) in Cornwall (from Xu & Thornton, 1985)

	Total As in soil ($\mu g\,g^{-1}$)	Lettuce ($\mu g\,g^{-1}$)	Onion ($\mu g\,g^{-1}$)	Beetroot ($\mu g\,g^{-1}$)	Carrot ($\mu g\,g^{-1}$)	Pea ($\mu g\,g^{-1}$)	Bean ($\mu g\,g^{-1}$)
Range	144–892	0.15–3.88	0.10–0.49	0.02–0.93	0.10–0.93	0.10–0.11	0.02–0.09
Geometric mean	322	0.85	0.20	0.17	0.21	0.04	0.04
No. of Samples	32	28	23	23	19	19	7

Fig. 4.5 Map showing the distribution of cadmium in stream sediment in England and Wales. (Part of the Wolfson Geochemical Atlas of England and Wales; Webb *et al.*, 1978.)

soils (0–5 cm) in Shipham had median concentrations of 91 ppm Cd and 7660 ppm Zn compared with 0.6 ppm Cd and 158 ppm Zn in North Petherton. In Shipham over 90% of soils contained in excess of 20 ppm Cd and 50% exceeded 60 ppm Cd (Fig. 4.6). There was a strong correlation between cadmium and zinc in these soils with a mean Zn:Cd ratio about 90:1.

The identification of such large amounts of cadmium in garden soils was unique in Britain and resulted in one of the most ambitious environmental health investigations ever mounted, involving national and local government, Westminster Hospital Medical School and Imperial College. These studies have been published collectively in 'The Shipham Report' (Morgan, 1988). It is not possible to describe the findings in detail but they may be summarized as follows:

1 The household garden soils greatly exceeded the levels of cadmium in polluted paddy soils associated with the well-documented 'itai-itai' disease in Japan.

2 Samples of housedust averaged 26 ppm cadmium and 2300 ppm zinc.

3 From studies of metals in locally grown vegetables and diets, an average intake for human beings of 200 µg cadmium per week was calculated, compared with the average intake in the United Kingdom of 140 µg cadmium per week. Individual intakes rarely exceeded the World Health Organisation's provisional tolerable weekly intake of 450–500 µg cadmium.

4 Health inventories and biochemical tests on 548 residents of Shipham and on 543 control subjects from a nearby uncontaminated village showed only slight differences attributable to cadmium.

Though elevated concentrations of cadmium were found in home-grown crops, the actual levels found reflected, only to a small degree, those in the soil, probably because the metal was not very mobile or plant-available due to the high soil pH and calcium carbonate content and other soil factors (Alloway *et al.*, 1988). In this unique situation it is also possible

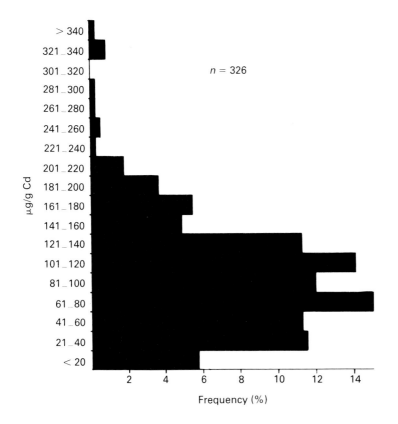

Fig. 4.6 Frequency distribution showing cadmium concentrations (µg g^{-1}) in Shipham garden soils (0–5 cm).

Brink, W.J. (eds), *Contaminated Soil*, pp. 759–768. Martinus Nijhoff, Dordrecht.

Bridges, E.M. (1987). *Surveying Derelict Land*. Clarendon Press, Oxford.

Cairney, T. (1986). Soil cover reclamation experience in Britain. In Assink, J.W. & van den Brink, W.J. (eds), *Contaminated Soil*, pp. 601–614. Martinus Nijhoff, Dordrecht.

Chamberlain, A.C. (1983). Fallout of lead and uptake by crops. *Atmospheric Environment* **17**, 693–706.

Chamberlain, A.C., Heard, M.J., Little, P., Newton, D., Wells, A.C. & Wiffen, R.D. (1978). *Investigations into Lead from Motor Vehicles*. Atomic Energy Research Establishment Report No. R9198, HMSO, London.

Collins, W.G. & Bush, P.W. (1969). The definition and classification of derelict land. *Journal of the Town Planning Institute* **55**, 111–115.

Craul, P.J. (1985). A description of urban soils and their desired characteristics. *Journal of Arboriculture* **11**, 330–339.

Culbard, E.B., Thornton, I., Watt, J., Wheatley, M., Moorcroft, S. & Thompson, M. (1988). Metal contamination in British suburban dusts and soils. *Journal of Environmental Quality* **17**, 226–234.

Davies, B.E. (1978). Plant-available lead and other metals in British garden soils. *The Science of the Total Environment* **9**, 243–262.

Davies, B.E. (1983). A graphical estimation of the normal lead content of some British soils. *Geoderma* **29**, 67–75.

Davies, B.E. (1986). Baseline survey of metals in Welsh soils. In Thornton, I. (ed.), *Proceedings First International Symposium on Geochemistry and Health*, pp. 45–51. Science Reviews, Northwood.

Davies, B.E. & Paveley, C.F. (1988). Baseline trace metal survey of Welsh soils with special reference to lead. In Thornton, I. (ed.), *Geochemistry and Health*, pp. 29–36. Science Reviews, Northwood.

Davies, B.E. & Thornton, I. (1990). *Environmental Pathways of Lead into Food*. A review, Lead Zinc Research Organisation, New York. 104 pp.

Davies, B.E., Conway, D. & Holt, S. (1979). Lead pollution of London soils: a potential restriction on their use for vegetable growing. *Journal of Agricultural Science (Cambridge)* **93**, 749–752.

Davies, D.J.A. (1987). An assessment of the exposure of young children to lead in the home environment. In Thornton, I. & Culbard, E. (eds), *Lead in the Home Environment*, pp. 189–196. Science Reviews, Northwood.

Davies, D.J.A. & Thornton, I. (1987). The influence of house age on lead levels in dusts and soils in Brighton, England. *Environmental Geochemistry and Health* **9**, 65–67.

Davies, D.J.A., Thornton, I., Watt, J.M., Culbard, E.B., Harvey, P.G., Delves, H.T., Sherlock, J.C., Smart, G.A., Thomas, J.F.A. & Quinn, M.J. (1987a). Relationship between blood lead and lead intake in two year old urban children in the UK. In Lundberg, S.E. & Hutchinson, T.C. (eds), *Heavy Metals in the Environment* vol. 2, pp. 203–205. New Orleans, CEP Consultants, Edinburgh.

Davies, D.J.A., Watt, J.M. & Thornton, I. (1987b). Lead levels in Birmingham dusts and soils. *The Science of the Total Environment* **67**, 177–185.

Department of the Environment (1982). *Pollution Report 15*. Central Directorate for Environmental Pollution, D.O.E., HMSO, London.

De Kreuk, J.F. (1986). Microbial decontamination of excavated soil. In Assink, J.W. & van den Brink, W.J. (eds), *Contaminated Soil*, pp. 669–678. Martinus Nijhoff, Dordrecht.

De Leer, E.W.B. (1986). Thermal methods developed in the Netherlands for the cleaning of contaminated soil. In Assink, J.W. & van den Brink, W.J. (eds), *Contaminated Soil*, pp. 645–654. Martinus Nijhoff, Dordrecht.

Downing, M.F. (1977). Survey information. In Hackett, B. (ed.), *Landscape Reclamation Practice*, pp. 17–36. IPC Science and Technology Press, London.

Haines, R. (1987). *Contaminated Sites in the West Midlands*. A Prospective Survey. JURUE, University of Aston, Birmingham.

Harrison, R.M. (1979). Toxic metals in street and household dusts. *The Science of the Total Environment* **11**, 89–97.

Hilberts, B., Eikelboorn, D.H., Verheul, J.H.A.M. & Heinis, F.S. (1986). *In situ* techniques. In Assink, J.W. & van den Brink, W.J. (eds), *Contaminated Soil*, pp. 679–978. Martinus Nijhoff, Dordrecht.

Holmgren, G.G.S., Meyer, M.W., Daniels, R.B., Kubota, J. & Chaney, R.L. (1983). Cadmium, zinc, copper and nickel in agricultural soils of the United States. *Agronomy Abstracts 1983*, 33.

Hutton, M., Wadge, A. & Milligan, P.J. (1988). Environmental levels of cadmium and lead in the vicinity of a major refuse incinerator. *Atmospheric Environment* **22**, 411–416.

Institute of Geological Sciences (1978). *Geochemical Atlas of Great Britain: Shetland Islands*. Institute of Geological Sciences, London.

Institute of Geological Sciences (1979). *Geochemical Atlas of Great Britain: Orkney Islands*. Institute of Geological Sciences, London.

Institute of Geological Sciences (1983). *Geochemical Atlas of Great Britain: Sutherland*. Institute of Geological Sciences, London.

Institute of Geological Sciences (1983/84). *Geochemical Atlas of Great Britain: South Orkney and Caithness*.

Institute of Geological Sciences, London.

Institute of Geological Sciences (1984). *Geochemical Atlas of Great Britain: Hebrides*. Institute of Geological Sciences, London.

Institute of Geological Sciences (1987). *Geochemical Atlas of Great Britain: Great Glen*. Institute of Geological Sciences, London.

Interdepartmental Committee for the Redevelopment of Contaminated Land. (1983). *Guidance on the assessment and redevelopment of contaminated land*. ICRCL Paper 59/83, Department of the Environment, London.

Interdepartmental Committee for the Redevelopment of Contaminated Land. (1987). *Guidance on the assessment and redevelopment of contaminated land*. ICRCL Paper 59/83, Second Edition. Department of the Environment, London.

Jessberger, H.L. (1986). Techniques for remedial action at waste disposal sites. In Assink, J.W. & van den Brink, W.J. (eds), *Contaminated Soil*, pp. 587–599.7 Martinus Nijhoff, Dordrecht.

Lepow, M.L., Bruckman, L., Gillette, M., Markowitz, S., Robino, R. & Kapish, J. (1975). Investigations into sources of lead in the environment of urban children. *Environmental Research* 40, 415–426.

Little, P. & Wiffen, R.D. (1978). Emission and deposition of lead from motor exhausts – II. Airborne concentration, particle size and deposition of lead near motorways. *Atmospheric Environment* 12, 1331–1341.

Ministry of Agriculture, Fisheries & Food (1982). *Survey of Arsenic in Food*. Food Surveillance Paper No. 8. HMSO, London.

McGrath, S.P. (1986). The range of metal concentrations in topsoils of England and Wales in relation to soil protection guidelines. In Hemphill, D.D. (ed.), *Trace Substances in Environmental Health XX*, pp. 242–252. University of Missouri, Columbia.

McGrath, S.P. (1987). Computerised quality control statistics and regional mapping of the concentrations of trace and major elements in the soil of England and Wales. *Soil Use and Management* 3, 31–38.

McGrath, S.P., Cunliffe, C.H. & Pope, A.J. (1986). Lead, zinc, cadmium, copper, nickel and chromium concentrations in the topsoils of England and Wales. In Thornton, I. (ed.), *Proceedings First International Symposium on Geochemistry and Health*, pp. 52–58. Science Review, Northwood.

McNeal, J.M. (1986). Regional scale geochemical mapping in the United States. In Thornton, I. (ed.), *Proceedings of the First International Symposium on Geochemistry and Health*, pp. 116–30. Science Reviews, Northwood.

Mitchell, R.L. (1964). Trace elements in soils. In Bear, F.E. (ed.), *Chemistry of the Soil*, pp. 320–368. Reinhold, New York.

Mitchell, R.L. (1974). Trace element problems in Scottish soils. *Netherlands Journal of Agricultural Science* 22, 295–304.

Moen, J.E.T., Cornet, J.P. & Evers, C.W.A. (1986). Soil protection and remedial actions: criteria for decision making and standardisation of requirements. In Assink, J.W. & van den Brink, W.J. (eds), *Contaminated Soil*, pp. 441–448. Martinus Nijhoff, Dordrecht.

Moorcroft, S.J., Watt, J., Wells, J., Thornton, I., Strehlow, C.D. & Barltrop, D. (1982). Composition of dusts and soils in an apparently uncontaminated rural village in southwest England – implications to human health. In Hemphill, D.D. (ed.), *Trace Substances in Environmental Health XVI*, pp. 1551–162. University of Missouri, Columbia.

Morgan, H. (1988). The Shipham Report. *The Science of the Total Environment* 7, 1–143.

Morgan, H. & Simms, D.L. (1988). Setting trigger concentrations for contaminated land. In Wolf, K., van den Brink, W.J. & Colen, F.J. (eds), *Contaminated Land '88*, Vol. I, pp. 327–337. Kluwer Academic, Dordrecht.

Moss, G.H. (1987). Wasting Europe's heritage – the need for soil protection. In Barth, H. & L'Hermite, P. (eds), *Scientific Basis for Soil Protection in the European Community*, pp. 17–28. Elsevier Applied Science, Barking.

Nriagu, J.O. & Pacyna, J.M. (1988). Quantitative assessment of worldwide contamination of air, water and soils by trace metals. *Nature, London* 333, 134–139.

Page, A.L. & Ganje, T.J. (1970). Accumulations of lead in soils for regions of high and low motor vehicle traffic density. *Environmental Science and Technology* 4, 140–142.

Plant, J.A. & Moore, P.J. (1979). Regional geochemical mapping and interpretation in Britain. *Philosophical Transactions of the Royal Society, London* B288, 95–112.

Plant, J.A. & Stevenson, A.G. (1986). Regional geochemistry and its role in epidemiological studies. In Mills, C.F., Bremner, I. & Chesters, J.K. (eds), *Trace Element Metabolism in Man and Animals V*, pp. 900–906. Rowett Research Institute, Aberdeen.

Plant, J.A. and Thornton, I. (1980). Regional geochemical mapping and health in the United Kingdom. *Journal of the Geological Society, London* 137, 575–586.

Plant, J.A. & Thornton, I. (1986). Geochemistry and health in the United Kingdom. In Thornton, I. (ed.), *Proceedings of the First International Symposium on Geochemistry and Health*, pp. 1–15. Science reviews, Northwood.

Reaves, G.A. & Berrow, M.L. (1984). Total lead concentrations in Scottish soils. *Geoderma* 32, 1–8.

Royal Commission on Environmental Pollution (1983). *Lead in the Environment*. 9th Report CMND 8852, HMSO, London.

Royal Commission on Environmental Pollution (1985).

Managing Waste: the Duty of Care. 11th Report CMND 9675, HMSO, London.

Rundle, S. & Duggan, M. (1980). *Report DE/SB/EDS/ R91*. Greater London Council, London.

Simms, D.L. & Beckett, M.J. (1987). Contaminated land: setting trigger concentrations. *The Science of the Total Environment* **65**, 121–134.

Smith, C.A. (1983). *The Distribution of Selenium in some Soils Developed on Silurian, Carboniferous and Cretaceous Systems in England and Wales*. PhD Thesis, University of London.

Smith, W.H. (1976). Lead contamination of the roadside ecosystem. *Journal of the Air Pollution Control Federation* **25**, 753–766.

Solomon, R.L. & Hartford, J.W. (1976). Lead and cadmium in dusts and soils in a small urban community. *Environmental Science and Technology* **10**, 773–777.

Thornton, I. (1980). Geochemical aspects of heavy metal pollution and agriculture in England and Wales. In *Inorganic Pollution and Agriculture*, pp. 105–125. MAFF Reference Book, 326. HMSO, London.

Thornton, I. (1983). Geochemistry applied to agriculture. In Thornton, I. (ed.), *Applied Environmental Geochemistry*, pp. 231–266. Academic Press, London.

Thornton, I. (1988). Metal content of soils and dusts. In Morgan, H. (ed.), The Shipham Report, *The Science of the Total Environment* **75**, 21–39.

Thornton, I. & Abrahams, P. (1984). Historical records of metal pollution in the environment. In Nriagu, J. (ed.). *Changing Metal Cycles and Human Health*, pp. 7–25. Springer-Verlag, Berlin.

Thornton, I. & Jones, T.H. (1984). Sources of lead and associated metals in vegetables grown in British urban soils: uptake from soil versus air deposition. In Hemphill, D.D. (ed.), *Trace Substances in Environmental Health XVIII*, pp. 303–310. University of Missouri, Columbia.

Thornton, I. & Webb, J.S. (1979). Geochemistry and health in the United Kingdom. *Philosophical Transactions of the Royal Society, London* **B288**, 151–168.

Thorton, I., John, S., Moorcroft, S. & Watt, J. (1980). Cadmium at Shipham – a unique example of environmental geochemistry and health. In Hemphill, D.D. (ed.), *Trace Substances in Environmental Health XVI*, pp. 27–37. University of Missouri, Columbia.

Thornton, I., Abrahams, P.W., Culbard, E., Rother, J.A.P. & Olson, B.H. (1986). The interaction between geochemical and pollutant metal sources in the environment: Implications for the community. In Thornton, I. & Howarth, R.J. (eds), *Applied Geochemistry in the 1980s*, pp. 270–308. Graham and Trotman, London.

Thornton, I., Culbard, E.B., Moorcroft, S., Watt, J., Wheatley, M., Thompson, M. & Thomas, J.F.A. (1985). Metals in urban dusts and soils. *Environmental Technology Letters* **6**, 137–145.

Wadge, A. & Hutton, M. (1987). The cadmium and lead content of suspended particulate matter emitted from a UK refuse incinerator. *The Science of the Total Environment* **67**, 91–95.

Webb, J.S. & Howarth, R.J. (1979). Regional geochemical mapping. *Philosophical Transactions of the Royal Society, London* **B288**, 81–93.

Webb, J.S., Nichol, I., Foster, R., Lowenstein, P.L. & Howarth, R.J. (1973). *Provisional Geochemical Atlas of Northern Ireland*. Applied Geochemistry Research Group, Imperial College of Science and Technology, London.

Webb, J.S., Thornton, I., Thompson, M., Howarth, R.J. & Lowenstein, P.L. (1978). *The Wolfson Geochemical Atlas of England and Wales*. Oxford University Press, Oxford.

Xu, J. & Thornton, I. (1985). Arsenic in garden soils and vegetable crops in Cornwall, England: Implications for human health. *Environmental Geochemistry and Health* **7**, 131–133.

5 Soil storage and handling

D.L. RIMMER

Introduction

The features which usually distinguish urban soils from their natural counterparts are the result of man's activity. In particular urban soils will often have been subjected to mixing, filling, disturbance or contamination (Craul, 1985). Thus at least part of the soil material is likely to have been either stored or handled or both at some time in its history. The effects of storage and handling on soil properties are the subject of this review.

Soil handling and storage associated with urban development is usually on a relatively small scale with sites extending over areas of the order of a few hectares. The handling and storage of soil is carried out on an *ad hoc* basis as it is not a matter of primary concern to the contractors involved. On the other hand, land restoration schemes associated with mineral extraction, such as for coal, gravel and ironstone, are often on a much larger scale with sites of tens or hundreds of hectares in extent. The storage and replacement of soil is an essential part of such schemes. Consequently the scientific literature on this subject is dominated by work relating to land restoration after mining, rather than urban development. If one bears in mind the differences in the two situations, such as the fact that for soils in urban areas the periods of storage are likely to be shorter but also more frequent, then it should be possible to transfer the findings from land restoration to the urban environment.

The effects of soil handling alone are primarily physical; but indirectly there are effects on the biological, microbiological and chemical status of soils. In most cases the effects of handling have not been measured separately from those arising from storage. Much of the literature dealing with the effects of storage considers the changing soil properties in soil storage heaps. However all stored soil will have been previously excavated and transported and this will also have had an effect. Consequently it is necessary to consider the combined effects of storage and handling. In so far as the soil is affected

by this treatment it is important to consider what happens after storage, to what extent there is recovery of the original status and at what rate that takes place.

The effects of soil handling

In a recent review of soil handling for quarry restoration, Ramsay (1986) emphasized that soil compaction and loss of structure are the major effects caused during the movement of soil. The simplest and most widely used measure of soil compaction in particular, and soil structure in general, is bulk density. Whilst not presenting any data specifically relating to soil handling on restoration sites, Ramsay (1986) reviewed the well-established literature relating to agricultural soils. Values of bulk density typical of undamaged mineral soils of different textural classes were quoted from Hausenbuiller (1972). These range from 1.00 $t\,m^{-3}$ for aggregated clay to 1.55 $t\,m^{-3}$ for sand. At high density values the soil has greatly reduced pore space and this leads to a reduction in rates of water movement and gas exchange, lower water-holding capacity and an increased resistance to root penetration (see Chapter 6). For example Veihmeyer & Hendrickson, in a classic paper published in 1948, showed that the critical bulk densities for root growth were 1.46 $t\,m^{-3}$ for clay soils and 1.75 $t\,m^{-3}$ for sandy soils. Above these values root growth is likely to be severely restricted.

King (1988) presented some measurements of bulk density for soil which had been restored progressively after opencast coal mining (Table 5.1). In progressive restoration there is no soil storage; the soil which is stripped at the 'front' of the site is then moved around to the area being filled and re-spread. There was a general increase in bulk density in all horizons of the restored soil compared to the undisturbed soil. The greatest increase was in the upper subsoil horizon, because soil tillage had decreased the compaction in the topsoil. Similar findings were reported by Reeve (1987) from a survey of the

Table 5.1 Bulk density of soil horizons from undisturbed and restored opencast land (King, 1988)

Undisturbed soil				
Horizon	Ap	Bg	BCg1	BCg2
Depth (m)	0–0.26	0.26–0.44	0.44–0.71	0.71–1.20
Bulk density (t m^{-3})	1.29	1.46	1.62	1.60
Restored land				
Horizon	A	B1	B2	B3
Depth (m)	0–0.21	0.21–0.51	0.51–0.82	0.82–1.20
Bulk density (t m^{-3})	1.38	1.64	1.68	1.68

physical condition of restored land in which the effects of different soil handling equipment were compared. Topsoil conditions were similar on the different treatments and depended on recent tillage. On the other hand subsoils differed greatly. Areas on which earthscrapers had operated had subsoils with greater strength than those restored using dumpers. Overall, Reeve (1987) concluded that the measurement most closely related to plant growth was the coarse porosity in the subsoil, which is related to permeability to water.

During soil handling the extent of damage and deterioration in soil physical conditions will depend on four factors. These are soil type, soil moisture content, soil handling process and the equipment used. Ramsay (1986) discusses each of these. Soils with a high sand content are often weakly structured such that the forces applied to them during handling lead to disaggregation and the formation of very dense structures of closely-packed sand grains. Such damage will occur at all moisture contents, but will be most severe when the soil is wet. Conversely soils with a higher clay content are resistant to damage when dry, but become susceptible at high moisture content (above the so-called 'plastic limit'). Soil strength is very much reduced for these soils when wet and they are likely to be damaged by a combination of compaction and smearing.

Ramsay identified lifting, transport and placement as the processes involved in soil handling. Soil damage during lifting results from a combination of compaction, caused by the passage of heavy machinery over the soil, and smearing, caused by wheelslip and the cutting action of the soil lifting equipment.

Because of their much higher ground pressure, earthscrapers can do more damage than tracked vehicles. However, earthscrapers have the advantage of being able to lift, transport and replace soil in a single operation. This makes them very efficient and they are used extensively on land restoration sites. On small urban development sites their use may be less justifiable. To minimize soil structural damage earthscrapers should only be used in dry conditions when soil strength is high. Transporting soil causes little damage to the soil being moved, but compaction will take place in soil which is being travelled over. During soil placement, compaction is again likely, particularly when earthscrapers are running over soil which has just been replaced.

The combined effects of handling and storage of soils

INTRODUCTION

Concern about the effects of soil disturbance during opencast coal mining first appeared in the early 1950s. A paper by Dougal (1950) highlighted a number of soil problems resulting from opencast coal mining on agricultural land. He identified: (i) consolidation of the topsoil; (ii) destruction of soil aggregates; (iii) disruption of systems of drainage; (iv) a number of associated physical and chemical changes which arose as a result of these first three effects, and finally (v) an apparent loss of soil fertility status. The latter was specifically measured as a reduction in both soil pH and plant available levels of the major plant nutrients nitrogen (N), potassium (K) and phosphorus (P).

This work was followed by a number of more detailed studies which concentrated on, for example, physical changes (Hunter & Currie, 1956) and microbiological effects (Barkworth & Bateson, 1964). A more general consideration of the effects of topsoil storage was published by O'Flanagan *et al.* (1963). They studied storage heaps at two opencast coal mining sites in Lancashire. In both cases the soil was stored for three years and was of sandy loam texture. Chemical, physical, bacteriological and biological measurements were made. Organic matter content (expressed as organic carbon) was measured before, during and after storage. The differences between samples taken before and after restoration were only very small and not thought to be significant (Table 5.2). The data suggest, however, that there is a reduction in organic matter during storage. The mixing of topsoil with subsoil during soil lifting is often responsible for such reductions rather than any process taking place within the storage heap itself. O'Flanagan *et al.* noted some contamination of the topsoil at site II, but very little at site I.

The small reductions in organic matter during storage were reflected in changes in the soil aggregate stability. Any appreciable loss of organic matter in a sandy loam soil might be expected to lead to a reduction in aggregate stability and an increased susceptibility to soil structural damage. However, samples taken at the end of the storage period had good aggregate stability as measured by the methods of Cooke & Williams (1961). The authors also noted that the orginal topsoils had 'a well-developed crumb structure' and that the replaced topsoil had 'maintained a moderately well-developed crumb . . . structure'. It seems therefore that at these sites the soil handling had not led to the loss of structure which is usually expected (see e.g. Ramsay, 1986). The reason for this is that these were coarse-textured soils with good structural stability due to the presence of moderately high levels of organic matter. Although no information is available, it must also be assumed that the soil-handling operations were carried out under favourable (i.e. dry) conditions.

Measurements of nitrate- and ammonium-nitrogen were also reported by O'Flanagan *et al.* (1963). Extractions were made on samples from the storage heaps before and after a period of incubation at 25°C (Table 5.3). These data clearly show the build-up of ammonium at depth within the storage heaps. This will arise because of the inability of nitrifying bacteria to function in the anaerobic conditions which will prevail within the heaps. When aerobic conditions are restored, on respreading the soil in the field, or as in this case, by incubating for three weeks in the laboratory, then the nitrifying bacteria can again convert ammonium to nitrate. This is clearly the case for site I but for site II only a small amount of nitrate production took place during incubation. For these samples there appeared to be a reduction in nitrifying bacteria during storage but subsequent sampling of replaced soil showed that the population was rapidly restored.

The bacteriological measurements showed that storage at both sites caused no difference in the numbers of anaerobic bacteria compared to the control. However, the numbers of aerobic bacteria declined in the stored soils especially at the lower depths. This reduction was rapidly reversed when the soils were restored. These bacteriological findings confirm the interpretation given above for the temporary reduction in nitrifying activity.

Soil handling in the construction of the storage

	Site I		Site II	
	Range	Mean	Range	Mean
Before disturbance	1.31–2.37	1.63	1.81–2.71	2.04
During storage (sampled at 60–90 cm)	1.21–1.34	–	1.47–1.64	–
After restoration		1.86		2.37

Table 5.2 Organic carbon percentage in stored soil at two sites in Lancashire (O'Flanagan *et al.*, 1963)

Table 5.3 Ammonium- and nitrate-N $(mg\,kg^{-1})$ in stored soil at two sites in Lancashire (O'Flanagan *et al.*, 1963)

	Site I		Site II	
	0–15 cm	> 60 cm	0–15 cm	> 60 cm
Before incubation:				
NH_4-N	6.1	39.1	7.6	60.9
NO_3-N	0.4	1.8	1.2	1.2
After incubation:				
NH_4-N	4.2	10.0	10.0	58.7
NO_3-N	6.8	31.1	5.4	17.7

heaps had a dramatic effect on the earthworm populations. Sampling carried out four months after their formation showed that the surviving population was about 3% of that estimated to be present in the undisturbed soils. This reduced population did recover somewhat during the storage period, although there were considerable fluctuations. It was expected that following restoration there would be recovery to normal field numbers especially if the land was under grass for a period of three or four years.

This work by O'Flanagan *et al.* covered all the major effects of soil handling and storage. In the following sections the literature on physical, chemical, microbiological and biological effects will be discussed separately.

PHYSICAL EFFECTS

In the earliest paper devoted to the structural changes taking place in stored soil, Hunter & Currie (1956) showed that there is some deterioration associated with the opencast coal mining of the 'heavier boulder-clay soils' in Northumberland and Durham. They compared worked and unworked soils and presented data for porosity and bulk density, permeability (of both undisturbed and repacked cores) and aggregate stability. The bulk density values for the worked soil were in the range $1.58–1.64\,t\,m^{-3}$ compared to $1.12–1.15\,t\,m^{-3}$ for the unworked soil. This is in agreement with the general observation of Dougal (1950) concerning consolidation of the topsoil. The values for worked soil were above the critical value of $1.46\,t\,m^{-3}$ given by Veihmeyer & Hendrickson (1948) above which root growth is likely to be severely restricted in clay soils. This compaction of the topsoil would be expected to lead

to slow water infiltration and movement and the permeability measurements confirmed this. Undisturbed cores of the worked soil had permeability in the range $0.022 \times 10^{-6}–1.12 \times 10^{-6}\,m\,s^{-1}$ compared to a range of $1.76 \times 10^{-6}–14.32 \times 10^{-6}\,m\,s^{-1}$ for the unworked soil.

Soil aggregate stability was also affected by the handling and storage associated with opencast coal working (Hunter & Currie, 1956). They made a number of measurements by different methods and found the 'dispersion ratio' of Middleton (1930) to be the most satisfactory. The higher the dispersion ratio the more structurally unstable is the soil. Their data for two sites are presented in Table 5.4. They explained the loss of structural stability by suggesting that the anaerobic conditions within the storage heaps would result in the loss of organic compounds involved in soil aggregation. In order to test this idea they measured organic matter content and found that not only was there less in worked soils compared to unworked soils but it was also different in 'character' (i.e. was less easily oxidized by hydrogen peroxide). In a further experiment they subjected unworked soils to either aerobic or anaerobic incubation in the laboratory and found higher dispersion ratio values in the soils which had been incubated in anaerobic conditions. The loss of soil structural stability in stored soils will make them more susceptible to damage during the handling associated with respreading.

The findings of Hunter & Currie (1956) disagree with those of O'Flanagan *et al.* (1963), who found no significant reduction in organic matter and little change in structural condition following soil storage, as discussed previously. The reason for this is probably due to the difference in soil texture. O'Flanagan

Sampling depth (cm)	Blagdon soil		Milkhope soil	
	Worked	Unworked	Worked	Unworked
0–7.5	22.1	10.9	28.0	12.9
7.5–15.0	37.6	8.7	26.6	11.9
15.0–22.5	43.7	11.1	24.6	13.2

Table 5.4 Soil aggregate stability measured by dispersion ratio method (Hunter & Currie, 1956)

et al. were dealing with sandy loam soils whereas Hunter and Currie worked with soils with a higher clay content. The possibility of soil damage during initial handling is much greater for clay-rich soils. This would intensify the anaerobic conditions during storage and lead to the structural deterioration reported by Hunter and Currie. Such severe conditions may well not have existed in the storage heaps studied by O'Flanagan *et al.*

A major study on the effects of soil storage associated with opencast mining was carried out in New Zealand (Ross & Cairns, 1981; McQueen & Ross, 1982; Widdowson *et al.*, 1982). The physical properties measured included: porosity, bulk density, soil strength, water-holding capacity, aggregate size and aggregate stability (McQueen & Ross, 1982). Two topsoil storage heaps were studied. One of these was of silty clay loam texture and the adjacent undisturbed topsoil was of higher sand content with a texture of clay loam/sandy loam. At the other location both stored and undisturbed soils had the same texture which was clay loam. The soil had been in these storage heaps in one case for six years with prior storage at another site for five years, and in the other case for ten years prior to sampling.

In general, the stored soils had higher bulk densities and lower porosities than adjacent soils, in agreement with previous studies. The degree of compaction was however less, with bulk density values in the range $1.13–1.45\,t\,m^{-3}$ which were considered to be satisfactory. In contrast, strength, measured as penetration resistance, was considerably increased and on one site was thought likely to cause problems with root growth. The lack of any consistent trends in these properties with depth suggested that the compaction arose during the handling operations rather than as a result of the overburden pressure in the storage heaps themselves.

McQueen & Ross (1982) also assessed the aera-tion status of the soils by examining samples taken from different depths in the storage heaps. The presence or absence of mottling was used as a diagnostic criterion. At both sites the boundary between aerobic and anaerobic soil occurred at approximately 2 m. Measurements of aggregate size distribution showed that the mean aggregate size increased with depth and was also greater in anaerobic soil. Finally they found a significantly lower aggregate stability in the stored soils than in the controls. When, however, the aggregate stability data were reanalysed to test statistically the difference between aerobic and anaerobic samples there was no significant difference.

A similar set of measurements were also made by Abdul-Kareem & McRae (1984) on 18 topsoil storage heaps of varying texture, length of storage and construction features. They reported the results for three heaps of contrasting texture: one 'sandy', one 'loamy' and one 'clayey'. They observed the visual boundary between aerobic and anaerobic soil at different depths within the three heaps. In the clayey soil it occurred at 0.3 m, for the loamy soil it was at about 1.3 m and in the sandy soil it was 2 m. They also found that aggregate stability was reduced in the stored soils compared to the undisturbed control samples. Ten samples were taken from each storage heap at increasing depth to 2 m for the sandy and loamy soils and to 1.5 m for the clayey soil. For the sandy soil despite all samples coming from the aerobic zone, the aggregate stability was greatly reduced in all ten samples. Only samples from the two lowest depths in the loamy soil showed appreciable reductions in aggregate stability. These were just above and straddling the boundary between the aerobic and anaerobic zones. In the clayey soil there was a small reduction in the aggregate stability in the two samples from the aerobic zone and greater reductions in the samples from lower depths where the material was anaerobic.

Abdul-Kareem & McRae (1984) also reported bulk density values for samples taken from near the surface of the loamy and clayey soil heaps. These ranged from 1.5 to $1.8\,t\,m^{-3}$ for the loamy soil and from 1.1 to $1.3\,t\,m^{-3}$ for the clayey soil. These values seem satisfactory for the clayey soil; but for the loamy soil may be sufficiently high to restrict root growth and water movement.

CHEMICAL EFFECTS

Dougal (1950) commented that storage of soil led to the development of acidity. Subsequent publications did not substantiate this claim. The results of pH measurements presented by Widdowson *et al.* (1982) for two storage heaps in New Zealand were very variable. For one there was a general trend of decreasing pH; in the other the trend was of increasing pH. It was concluded that storage had caused minimal effects on the chemical properties which they had measured (pH, cation exchange capacity, exchangeable cations, organic carbon and nitrogen, available phosphorus (P) and P retention). Abdul-Kareem & McRae (1984) also measured pH and concluded that it increased, although the effect was marked in only one of the three storage heaps for which they presented data. Ross & Cairns (1981) also reported increased pH values at depth. Both groups attributed the pH increases to the accumulation of ammonium ions under the anaerobic conditions in the heaps.

It is these anaerobic conditions which are responsible for all of the significant changes in the chemical properties of stored soil. Abdul-Kareem & McRae (1984) measured extractable manganese and ferrous iron in the three storage heaps of different texture which they studied. Extremely large increases in ferrous iron concentrations were found in the anaerobic zone of the clayey soil. The effect with manganese was smaller for that soil but was quite marked for the sandy soil. These effects are only temporary; on respreading the soil and with the return of aerobic conditions the extractable iron and manganese levels will return rapidly to normal.

Many workers (O'Flanagan *et al.*, 1963; Ross & Cairns, 1981; Abdul-Kareem & McRae, 1984) have reported large increases in ammonium ion concentration (Table 5.3). Again this accumulation is only transitory. With the return of aerobic conditions the ammonium is converted to nitrate by the nitrifying bacteria (Table 5.3). The origin of the released ammonium ions is the soil organic matter, which continues to be broken down microbially during soil storage. Decreases in organic matter content have also been widely reported (Hunter & Currie, 1956; Abdul-Kareem & McRae, 1984). In contrast Widdowson *et al.* (1982) found either no change in organic matter or increasing amounts with depth in the storage heaps which they studied. They attributed any variation in organic matter and other associated properties to variability in the nature of the soils when the storage heaps were originally constructed. Such problems of heterogeneity clearly make interpretation of the data rather difficult in these studies.

MICROBIOLOGICAL EFFECTS

This area of investigation has received considerable attention, particularly from workers in North America. Various measurements have been made, including microbial biomass, and estimates of both bacterial and fungal populations, especially vesicular–arbuscular mycorrhizal (VAM) populations.

Most workers have reported a decrease in overall microbial biomass with increasing depth in soil storage heaps (Ross & Cairns, 1981; Abdul-Kareem & McRae, 1984; Harris & Birch, 1987). However, Visser *et al.* (1984a) reported that the only significant decrease in microbial biomass occurred in the surface soil of the storage heap which they studied. Over the three-year period of their study no significant difference was found between the microbial biomass at depth within the storage heap and that in the undisturbed control soil. The explanation of this disagreement could lie in the fact that a number of different methods of assessing microbial biomass were used by these workers. Ross & Cairns (1981) and Abdul-Kareem & McRae (1984) both used the chloroform fumigation method of Jenkinson & Powlson (1976); whereas Harris & Birch (1987) used an ATP assay and Visser *et al.* (1984a) used the glucose amendment method of Anderson & Domsch (1978). Visser *et al.* (1984a) explained the decreased microbial biomass in the surface of the storage heap as resulting from dilution of topsoil with subsoil material of lower organic matter content and also the effects of weathering at the surface.

The magnitude of these reported decreases in microbial biomass was large. Harris & Birch (1987) found that for the deepest sample in a four-year-old store only 15% of the total microbial biomass of the control was present. Ross & Cairns (1981) measured mineral-N flush as an estimate of microbial biomass and reported the following values for a ten-year-old storage heap: control soil (0–20 cm), 101 mg N kg^{-1} soil; stored soil (0–20 cm), 86 mg N kg^{-1} soil; stored soil (200 cm), 13 mg N kg^{-1} soil.

Visser *et al.* (1984a) also measured microbial respiration which is a measure of microbial activity. They found that storage had very little effect on this although they discovered that the stored soils were slower to respond to the glucose additions used in the Anderson & Domsch method of assessing microbial biomass than were undisturbed soils. They suggested that this response may be a more sensitive indicator of changes in soil microbiology than either respiration or biomass.

Time of storage will be an important variable in determining the extent of changes to the microbial populations in stored soil. Harris & Birch (1987) concluded from their studies that there was a rapid initial effect within the first few months then more gradual change thereafter. This was certainly the case with their ATP assays of microbial activity. Subsequent work has shown that some of the changes are extremely rapid with measurable decreases in microbial populations in less than one day (Harris, pers. comm.).

Studies of bacterial populations in stored soil were first published in 1964 when Barkworth and Bateson reported the results of analyses of 27 soil storage heaps at a variety of coal, ironstone and gravel workings. They concluded that bacteria die very slowly; at the oldest site analysed (14 years old) they found a reasonably high population of bacteria representative of the normal soil flora. More recent work by Harris & Birch (1987) showed that the deepest sample from a four-year-old site had only 18% of the total bacterial numbers of the control site. The effect of time of storage was also investigated by these authors. They found that bacterial numbers initially increased at all depths in the storage heaps before declining after a period of a few months. The reason for the initial increase is the additional substrate resulting from the death of other components of the microbial population.

The most obvious effect of storage on the bacterial population is the reduction in the nitrifier population discussed earlier in relation to the work of O'Flanagan *et al.* (1963) and others. The nitrifier population requires aerobic soil conditions and is adversely affected in the anaerobic zones at depth within the storage heaps.

General assessments of fungal groups were carried out by Harris & Birch (1987) and compared to their findings for bacteria. The effect of the soil conditions at depth was more marked for the fungal groups than for bacteria, there being a greater decrease in numbers of the fungi. In addition, the decrease was immediate for the fungi unlike the bacteria which increased initially before declining.

The considerable interest shown in vesicular–arbuscular mycorrhizae (VAM) arose because of the requirement of certain plants to establish a mycorrhizal association in order to improve their uptake of phosphorus. In re-establishing vegetation on soil which had been stored such plants would require there to be a satisfactory infectivity if they were to survive (Reeves *et al.*, 1981). In an early study Rives *et al.* (1980) found that there was a substantial decrease in viable inoculum in a soil stored for three years compared to the control. In a follow-up study on the same soils, Gould & Liberta (1981) found that the inoculation potential as measured by a bioassay of maize roots was unchanged in the undisturbed soil 15 months later, but in the stored soil there had been a further decrease, only 50.4% of the maize roots were mycorrhizal compared to 81.7% previously. Similar results were reported by Abdul-Kareem & McRae (1984) who found that infection was six to ten times greater in undisturbed soil than in sandy soil which had been stored for seven years. In a bioassay using lettuce roots the stored soil caused 67% infection compared to 79% for the undisturbed soil.

This reduction in infectivity was one reason given by Visser *et al.* (1984b) for the delay in infection in grasses grown in stored soil compared to controls. However it was also found that there was a shift in the dominant species of VAM fungus involved in infection from *Glomus fasciculatum* in the control soil to *G. mosseae* in the stored soil. This latter shift could also explain the differences in rate of infection between the two soils.

In the semi-arid environment of the western

Table 5.5 Infectivity determined by maize bioassay (% VAM infection) (Harris *et al.*, 1987)

Depth (m)	3-month old heap	4-year old heap
0	1.1	0.5
0.25	0.6	0.2
0.5	0.8	0
1.0	0.2	0.3
2.0	0	0
3.0	0	0

United States, Miller *et al.* (1985) found moisture status was an important determinant of VAM propagule survival. This led them to conclude that two populations probably existed adapted to moisture potentials above and below the critical value of 2 MPa. Under dry conditions length of storage was not an important factor in propagule survival, whereas under wet conditions it was. They therefore suggested that, in arid or semi-arid environments, soil storage heaps should be large and only constructed during dry periods.

Harris *et al.* (1987) studied the VAM populations of two storage heaps, one three months old and the other four years old. The infectivity of the older heap was less than that of the younger one (Table 5.5). The inoculum appeared to be associated with infected root fragments and with fresh roots rather than with spores although the latter may have been significant in the younger store. Infectivity decreased with depth within the heaps and they concluded that below 1 m propagules appeared to be non-viable (Table 5.5).

BIOLOGICAL EFFECTS

The effects of soil storage on soil fauna have not been as extensively studied as those on soil micro-organisms. Two aspects which have received attention are earthworm populations and seed viability.

As previously discussed, O'Flanagan *et al.* (1963) observed reductions in the earthworm populations in soil storage heaps. Abdul-Kareem & McRae (1984) also measured earthworm populations in storage heaps and in adjacent unworked soils. They found that earthworm numbers and earthworm biomass (g m^{-2}) both decreased in the storage heaps. For example, in the clayey Tonbridge soil, which

had been stored for 18 months, the control had a population of 141 earthworms per square metre compared to only 20 m^{-2} in the storage heap. In an earlier study Standen *et al.* (1982) reported a survey of earthworm populations in both opencast reclamation sites and colliery spoil heaps; similar decreases in earthworm numbers and biomass were found in a three-year-old storage heap. These workers also followed the population development after site restoration and found that the population was zero after one year but recovered somewhat in the second year. They also reported differences between the dominant groups of species found in the storage heaps compared to those in subsequently restored land. They explained the presence of earthworms in the storage heap compared to their absence subsequently as being the result of only one disturbance in the former case whereas in the latter case the soil had been disturbed twice. Rushton (1986) has reported similar findings on restored, opencast coal-mining land. Again he could find no earthworms on a site which had been reclaimed for only one year.

In some restoration schemes, it is required to restore natural or semi-natural vegetation. In such cases, the seed bank in the original soil can provide the basis for the vegetation on the restored site. However, in order for this to be successful, the seeds must remain viable during soil storage. Dickie *et al.* (1988) took samples from two topsoil stores, three months and four years old. The viable seed populations were measured by incubation (Table 5.6). Marked differences were found both with age of storage heap and with depth of sample. Overall, there was a decrease in viable seed numbers with depth which was particularly apparent in the older store. In interpreting the data presented in Table 5.6, the authors warned that caution must be exercised because of heterogeneity created during the formation of the heaps when mixing of subsoil with

Table 5.6 Total numbers of seedlings emerging from three depths in two storage heaps (Dickie *et al.*, 1988)

Age of store	Depth (m)		
	0	1	2
3 months	117	179	54
4 years	130	10	22

topsoil can occur. This mixing effect was particularly marked in the work of Tacey & Glossop (1980). They assessed three topsoil handling techniques at mined sites in Western Australia and measured seed germination after soil placement. The three techniques were replacement of 40 cm of stored topsoil, 40 cm of directly placed topsoil, and double-stripped topsoil. The latter treatment involved the placement of 5 cm of freshly stripped topsoil over a layer of stored soil. They found no significant difference between the number of seeds germinating in the stored topsoil compared to the directly placed topsoil, but very much increased germination in the double-stripped treatment. They showed that most of the seeds were in the top 2 cm of the undisturbed soil and that stripping 40 cm and mixing was sufficient to reduce subsequent germination. However, careful stripping of a very shallow layer (5 cm) preserved most of the seed bank and led to a high rate of germination.

Recovery of soil properties after soil replacement

The rate at which the soil recovers from the changes caused by storage and handling is as important as the magnitude of the changes themselves. Any changes which are reversed within a few weeks or months of soil replacement are of little significance. Changes of this type include chemical effects resulting from the development of anaerobic conditions within the storage heap. Abdul-Kareem & McRae (1984) reported that the concentration of ferrous iron in a sample taken from the centre of a heap during re-spreading was high ($1028 \, mg \, g^{-1}$), but two weeks later had decreased to $124 \, mg \, g^{-1}$. Similar behaviour would be expected for manganese.

There is also evidence that the nitrifying bacteria rapidly recover after soil replacement (see the discussion of the data in Table 5.3). However, it was observed by Ross & Cairns (1981) that recovery of microbial biomass in soil from the bottom of the stockpiles (where the anaerobic conditions would have been most severe) could be slow. These authors also found that the rate of recovery of nitrifying activity varied considerably from sample to sample and suggested that further data were needed in order to ascertain recovery rates under field conditions.

A more substantial research effort has been undertaken in the investigation of VAM fungi including their recovery after soil replacement. Allen & Allen (1980) studied six reclaimed mine sites at which topsoil had been mixed with subsoil. They measured percentage root infection by VAM fungi and soil spores. On two- and three-year-old sites both of these measures were reduced to about 50% or less of those on undisturbed sites. Stark & Redente (1987) showed that the productivity of stored soil would depend on the species being grown. They found that production was reduced in previously stored topsoil if the species depended on mycorrhizal associations. In general, work in the United States has shown that recovery of mycorrhizal populations takes many years. However, Warner (1983) working in England reported that the re-establishment of VAM fungi occurred after only eight to 20 months depending on cropping practices. These differences between the situation in England and the United States are most likely related to differences in climate and the associated vegetational succession.

The recovery of earthworm populations following soil reinstatement has been studied by a number of workers. Standen *et al.* (1982) studied three sites aged one, two and 11 years since restoration. Earthworm numbers were respectively 0, 11 and $33 \, m^{-2}$. On an adjacent undisturbed, but recently ploughed and re-seeded, area the number was $22 \, m^{-2}$. Rushton (1986) sampled a newly reclaimed site one year after reclamation and no earthworms were found. Samples taken at the same site over the following two-year period had a mean population density of $20 \, m^{-2}$. These studies seem to bear out the general observation of Chapman (quoted by Brook & Bates, 1960) that earthworm populations recovered to normal within four years of restoration.

The build-up of organic matter, depleted during soil storage, will be a slow process, which will depend on the soil management practices. The rate of recovery will be maximized by having grass and by adding organic matter for example as farmyard manure or sewage sludge. Such practices will also hasten the recovery of earthworm populations (Rushton, 1986) and soil structure (Brook & Bates, 1960). Damage to soil structure will be particularly slow to rectify. Whilst it is possible to control physical conditions in the surface soil by means of appropriate cultivations (Reeve, 1987), conditions in the subsoil are likely to be both poorer and less easy to rectify. Information relating to changing soil physical conditions on restored land is particularly scarce. King (1988) has shown the advantages of installing a drainage system at the time of restoration. Although most of his measurements were made in the first two years after installation of the drainage system, there was improved structure in the topsoil of the drained area compared to the undrained. This was thought to have resulted from improved root development and earthworm activity in the better-drained soil.

Discussion and conclusions

Damage to soil structure during soil handling, which leads to restored land being consolidated, depends on a combination of the handling methods (the equipment and how it is used) and the soil conditions (particularly moisture content and initial structural state). Our understanding of soil mechanics, and the factors affecting soil strength, is sufficiently good to be able to predict the extent of soil damage under known conditions and also to be able to recommend methods of soil handling which can minimize such structural damage (Ramsay, 1986).

The effects of storage on soil structure appear to be relatively minor and related to the loss of organic matter, which reduces aggregate stability. The importance of this is uncertain, although it could increase susceptibility to damage during subsequent re-spreading operations, especially if the soil was wet.

It is clear that soil structure is damaged by either storage or handling or both and the important unanswered question is how rapidly does the structure recover? There is little information, apart from some field observations (King, 1988). Both field and laboratory research are required to determine the factors affecting structure recovery such as the influence of particle size distribution, clay mineralogy, drying and freezing cycles.

Chemical changes induced by soil storage are relatively unimportant, readily reversible and, where necessary, easily rectified by fertilizer application (Ministry of Agriculture, Fisheries and Food, 1978). The changes in microbial populations are now reasonably well documented. Some doubt remains, however, as to the rate of recovery of nitrifying bacteria. Our current knowledge about the effects of storage and handling on soil fauna is limited to earthworms, for which the effects of handling are more important than those of soil storage. There may be a case for a wider survey covering, for example, mites, collembola and other arthropods. The environmental conditions within soil storage heaps are such that loss of seed viability is inevitable. To overcome this problem, Dickie *et al.* (1988) recommended that, where soil storage for more than a few months was necessary, viable seeds should be removed from the site before soil stripping and stored under optimum conditions until required.

References

Abdul-Kareem, A.W. & McRae, S.G. (1984). The effects on topsoil of long-term storage in stockpiles. *Plant and Soil* **76**, 357–363.

Allen, E.B. & Allen, M.F. (1980). Natural re-establishment of vesicular–arbuscular mycorrhizae following stripmine reclamation in Wyoming. *Journal of Applied Ecology* **17**, 139–147.

Anderson, J.P.E. & Domsch, K.H. (1978). A physiological method for the quantitative measurement of microbial biomass in soil. *Soil Biology and Biochemistry* **10**, 215–221.

Barkworth, H. & Bateson, M. (1964). An investigation into the bacteriology of topsoil dumps. *Plant and Soil* **21**, 345–353.

Brook, D.S. & Bates, F. (1960). Grassland in the restoration of opencast coal sites in Yorkshire. *Journal of the British Grassland Society* **15**, 116–123.

Cooke, G.W. & Williams, R.J.B. (1961). Some effects of farmyard manure and of grass residues on soil structure. *Soil Science* **92**, 30–39.

Craul, P.J. (1985). A description of urban soils and their desired characteristics. *Journal of Arboriculture* **11**, 330–339.

Dickie, J.B., Gajjar, K.H., Birch, P. & Harris, J.A. (1988). The survival of viable seeds in stored topsoil from opencast coal workings and its implication for site restoration. *Biological Conservation* **43**, 257–265.

Dougal, B.M. (1950). The effects of opencast coal mining on agricultural land. *Journal of the Science of Food and Agriculture* **1**, 225–229.

Gould, A.B. & Liberta, A.E. (1981). Effects of topsoil storage during surface mining on the viability of vesicular–arbuscular mycorrhiza. *Mycologia* **73**, 914–922.

Harris, J.A. & Birch P. (1987). The effects on topsoil of storage during opencast mining operations. *Journal of the Science of Food and Agriculture* **40**, 220–221.

Harris, J.A., Hunter, D., Birch, P. & Short, K.C. (1987). Vesicular–arbuscular mycorrhizal populations in stored topsoil. *Transactions of the British Mycological Society* **89**, 600–602.

Hausenbuiller, R.L. (1972). *Soil Science: Principles and Practice.* William C. Braun, Iowa.

Hunter, F. & Currie, J.A. (1956). Structural changes during bulk soil storage. *Journal of Soil Science* **7**, 75–80.

Jenkinson, D.S. & Powlson, D.S. (1976). The effects of biocidal treatments on metabolism in soil. V. A method for measuring soil biomass. *Soil Biology and Biochemistry* **8**, 209–213.

King, J.A. (1988). Some physical features of soil after opencast mining. *Soil Use and Management* **4**, 23–30.

McQueen, D.J. & Ross, C.W. (1982). Effects of stockpiling topsoils associated with opencast mining. 2. Physical properties. *New Zealand Journal of Science* **25**, 295–302.

Middleton, H.E. (1930). Properties of soils which influence soil erosion. *USDA Technical Bulletin* **178**, pp. 16.

Miller, R.M., Carnes, B.A. & Moorman, T.B. (1985). Factors influencing survival of vesicular–arbuscular mycorrhiza propagules during topsoil storage. *Journal of Applied Ecology* **22**, 259–266.

Ministry of Agriculture, Fisheries and Food. (1978). *Farming Restored Opencast Land.* ADAS Advisory Leaflet No. 510. HMSO, London.

O'Flanagan, N.C., Walker, G.J., Waller, W.M. & Murdoch, G. (1963). Changes taking place in topsoil stored in heaps on opencast sites. *NAAS Quarterly Review* **62**, 85–92.

Ramsay, W.J.H. (1986). Bulk soil handling for quarry restoration. *Soil Use and Management* **2**, 30–39.

Reeve, M.J. (1987). Variations in the physical condition of land restored to agriculture. *Journal of the Science of Food and Agriculture* **40**, 226–227.

Reeves, F.B., Schmidt, S., Sabaloni, J., Schwab, S. & Redente, E. (1981). Changes of mycorrhizal inoculum potential in stockpiled topsoil. *Phytopathology* **71**, 1006.

Rives, C.S., Bajwi, M.I., Liberta, A.E. & Miller, R.M. (1980). Effects of topsoil storage during surface mining on the viability of VA mycorrhiza. *Soil Science* **129**, 253–257.

Ross, D.J. & Cairns, A. (1981). Nitrogen availability and microbial biomass in stockpiled topsoils in Southland. *New Zealand Journal of Science* **24**, 137–143.

Rushton, S.P. (1986). Development of earthworm populations on pasture land reclaimed from opencast coal mining. *Pedobiologia* **29**, 27–32.

Standen, V., Stead, G.B. & Dunning, A. (1982). Lumbricid populations in opencast reclamation sites and colliery spoil heaps in County Durham, UK. *Pedobiologia* **24**, 57–64.

Stark, J.M. & Redente, E.F. (1987). Production potential of stockpiled topsoil. *Soil Science* **144**, 72–76.

Tacey, W.H. & Glossop, B.L. (1980). Assessment of topsoil handling techniques for rehabilitation of sites mined for bauxite within the jarrah forest of Western Australia. *Journal of Applied Ecology* **17**, 195–201.

Veihmeyer, F.J. & Hendrickson, A.H. (1948). Soil density and root penetration. *Soil Science* **65**, 487–493.

Visser, S., Fujikawa, J., Griffiths, C.L. & Parkinson, D. (1984a). Effect of topsoil storage on microbial activity, primary production and decomposition potential. *Plant and Soil* **82**, 41–50.

Visser, S., Griffiths, C.L. & Parkinson, D. (1984b). Topsoil storage effects on primary production and rates of vesicular-arbuscular mycorrhizal development in *Agropyron trachycaulum. Plant and Soil* **82**, 51–60.

Warner, A. (1983). Re-establishment of indigenous vesicular–arbuscular mycorrhizal fungi after topsoil storage. *pyron trachycaulum. Plant and Soil* **82**, 51–60.

Widdowson, J.P., Gibson, E.J. & Healy, W.B. (1982). Effects of stockpiling topsoils associated with opencast mining. 1. Chemical properties and the growth of ryegrass and white clover. *New Zealand Journal of Science* **25**, 287–294.

6 Physical properties of soils in urban areas

C.E. MULLINS

Soils of urban areas are under increased environmental pressure because of the intensive uses to which they may be put. The major division between such uses is between that for amenity purposes (e.g. parks, gardens and playing fields) and that for the disposal of waste products. The other use of urban soil, for the foundations of buildings and roads, forms the basis of the discipline of soil mechanics but will not be discussed here.

This chapter is divided into two parts corresponding to its two aims. The first part deals with soil physical properties and processes relevant to urban land use, and the second part reviews some of the specialist literature relevant to particular types of urban land use. No attempt has been made to provide a comprehensive review of such an extensive and widely scattered literature; rather, the scientific principles are outlined, their application in the context of the use of soils in urban areas is discussed, and some gaps and weak points in our knowledge are identified.

What is 'urban soil'?

Despite the classification of soils in urban areas (see Chapter 2), it is worth remembering that the soil may well be heterogeneous and contain the remains of many types of activity. Most commonly this can take the form of building rubble or other forms of waste material which lie buried and hidden from sight. Even in parkland there may be the forgotten remains of previous human activity such as buried Roman or medieval roads or buildings, etc. In areas where there is a high earthworm population, such as uncompacted land under grass, these large dense objects tend to sink into the soil at a surprising rate. Darwin (1882) observed the rate of sinkage and burial of stone slabs in a local graveyard as a result of the undermining action of earthworms and calculated that the top 5 cm of soil passed through the intestines of earthworms once every ten years. He also observed that such objects only sink to the depth of earthworm activity (c.0.5–1 m). Such features are unlikely to be registered on soil maps although they may exert a considerable influence on the pattern of water flowing through the soil and on the use of the soil as a medium for plant growth. A detailed soil survey may then be necessary.

In built-up areas the 'soil' can often contain interleaved strata each of which contains more or less mixed layers of building rubble, rubbish and soil (which itself may have been brought in from elsewhere). One glance at the soil profiles revealed during the excavations for urban archaeology or building foundations is sufficient to confirm this point. Urban gardens may consist of a subsoil containing scattered patches of rubble, all of which has been run over and compacted by heavy machinery, on top of which is placed a layer (c. 0.2–0.5 m) of 'topsoil'.

It is thus clear that it is safest to consider 'urban soil' as an artificial material that has some of the physical properties of soils in their natural or agricultural environment, and the word 'soil' as used in this chapter will encompass such material. In this respect, the literature on soil restoration (e.g. after opencast mining or gravel extraction), where the thicknesses and treatments of the various replaced soil layers is often documented, clearly represents a valuable reference source of relevance to the urban situation. Similarly, a useful source of references relating to physical properties relevant to revegetation is to be found in papers on the revegetation of spoil heaps.

In an agricultural context 'topsoil' refers to that layer of soil which is regularly disturbed and often inverted by cultivation operations. Subsoil is the relatively less disturbed soil underneath. This terminology is imprecise and, in an urban context, its particular use will only be apparent from the context in which the words are used. Thus, in the context of land restoration, 'topsoil' is commonly used to refer to dark-coloured soil. It will often, though not always, have originated from the top 30 cm of a natural soil and then have been placed with minimal compaction as a surface layer. Subsoil is then the layer of soil underneath this that originates from a greater depth and is likely to have been compacted when the topsoil was replaced.

Sustainable use and soil damage

The physical changes that occur in soils in urban areas depend upon the intrinsic properties of the soil (e.g. particle size distribution), and the type of treatment or use to which it is subject (e.g. vegetation type, soil water regime and amount of treading). Apart from particle size distribution, most important soil physical properties (e.g. porosity, erodibility, structural stability, saturated hydraulic conductivity, structure and rootability) depend on the way in which the soil is used. Soil with a particular set of properties may be acceptable for one use but not for another more intensive use. Thus, soil damage has no unique definition because it depends upon the intended use of the soil. However, an operational definition is simply that a soil is unacceptably damaged when its intended use is no longer permanently sustainable. Where soil is vulnerable to damage, its management for sustainable use then has to be a compromise between two or more conflicting sets of factors.

Changes in the management or use of a soil can often lead to progressive changes in soil physical properties resulting in a vicious spiral of degradation. Playing fields provide an example of this point. Their repeated use in wet conditions may ultimately lead to unacceptable soil damage which renders the land unplayable (Fig. 6.1). These progressive changes may only become evident over a period

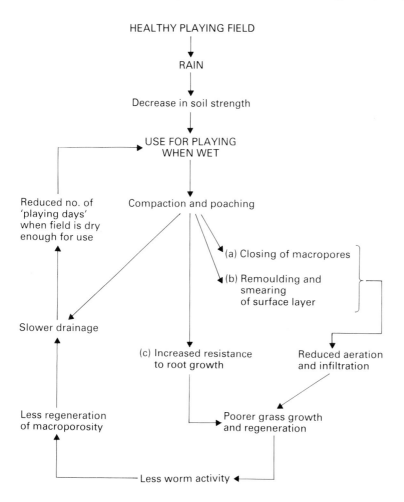

Fig. 6.1 Spiral of soil degradation resulting from overuse of playing fields under wet conditions.

Table 6.1 Categorization of soil physical properties

Set of properties	Notes	Requirements (see Table 6.2)
Particle size distribution	Particle size class/texture is a way of simplifying the detailed results of size analysis but omits some information	All
Phase relations	Density of solids, γ_s [often in the range $2.65–2.75\,t\,m^{-3}$], bulk density, γ_d, porosity, n [alternative measures of the degree of packing, $n = [1 - (\gamma_d/\gamma_s)]$] gravimetric water content, m volumetric water content, θ air porosity, $n_a\,[= n - \theta]$	All
Matric potential Ψ_m	Determines availability of water to plants. Hydraulic potential, Ψ_h is merely Ψ_m adjusted to some reference depth	C
Water release characteristic $\theta[\Psi_m]$	Fixed points on this characteristic are used to estimate: field capacity, $\theta[-5\,kPa]$; wilting point, $\theta[-1.5\,MPa]$; available water capacity, $\theta[-5\,kPa] - \theta[-1.5\,MPa]$; and air capacity, $n_a\,[-5\,kPa] \simeq \theta[0] - \theta[-5\,kPa]$	C
Structure	Qualitative description of peds, pores and crack geometry and spacing within soil available in Soil Survey manuals	A, C
Structural stability	Many tests have been proposed but are of limited practical applicability	D
Water transport properties	Infiltration depends on surface and near surface conditions. Below the water table, flux density, $q = -k\,(d\Psi_h/dz)$ where k is saturated hydraulic conductivity, and $(d\Psi_h/dz)$ is the gradient in hydraulic potential. k depends on structure (i.e. size and connectivity) of pores or on pore size distribution in structureless soil	A, C, D
Air transport properties	Diffusion coefficient, D, depends on n_a and pore connectivity	C
Soil strength	Shear and tensile strength and bulk compressibility are all relevant properties. Penetration resistance is a popular strength measurement which can be related to both load bearing capacity and, in a structureless soil, to root growth. Strength varies with n, θ and Ψ_m	B, C, F

of months or even years and, if soil deterioration can be identified at an early stage, appropriate remedial action can be taken. However, where damage has proceeded to the point that there are obvious signs of soil deterioration (e.g. poor growth of vegetation, surface runoff and ponding, or soil erosion), ameliorating the soil physical properties and vegetation is a major task whose cost may be prohibitive. Thus the experience of a skilled observer and/or monitoring of soil physical conditions can provide an early warning that the soil may

not withstand further intensive use, or that use should be restricted to certain periods. It is therefore important to know: (i) what soil physical properties are relevant to any given type of land use; (ii) for any given soil type, what range of values is to be expected for each physical property; (iii) how these properties can be measured and whether quick qualitative estimates for them are available; and (iv) how the sensitivity of these properties to the impact of different types of use varies with soil type. This chapter discusses these questions.

Soil physical properties

The most fundamental of soil physical properties is the particle size distribution of the soil since, unlike other properties, it is unaffected by the way in which the soil is managed (e.g. other physical properties can be changed by compaction). For this reason, particle size analysis can be performed in the laboratory on disturbed samples of soil, whereas almost all other physical properties have to be measured either *in situ* or on undisturbed samples brought back to the laboratory.

The pores or spaces between particles provide the pathways through which movement of soil water and soil air occurs. The phase relationships characterize the relative proportions of solid, liquid and gas in the soil and hence describe the amount of pore space and the proportions of it that are occupied by air and water. As water is removed from soil, the water remaining becomes progressively more difficult to extract. The relationship between water content and ease of extraction is described by the water release characteristic which also indicates the amount of plant available water that a soil is able to hold. Matric potential (Ψ_m) is a measure of the ease with which soil water can be extracted (e.g. by plant roots) whilst a related property, hydraulic potential, can be used to indicate the direction of soil water movement. Both potentials can easily be measured with tensiometers in moist soils ($\Psi_m > -80\,\text{kPa}$).

In soils containing worm channels, structural cracks and other macropores ($>60\,\mu\text{m}$ diameter), these are often the dominant pathways for the infiltration of water, drainage and soil aeration. In dense soils, structural cracks and macropores may also form the dominant pathway for root growth. Some soils are structurally unstable and may cap after heavy rainfall leading to runoff, possible erosion and, when dry, the cap may harden into a crust that prevents seedling emergence. Thus both soil structure and its stability are relevant in an urban context. Transport properties like the diffusion coefficient and hydraulic conductivity determine the ease of flow of gas and liquid in the soil and are themselves dependent on pore size distribution and the extent to which pores interconnect. Finally, soil strength, both in terms of the soil's ability to withstand loads (such as the frequent treading of playing fields) and in terms of ease of plant root

growth, may be characterized in a number of ways (Table 6.1).

Table 6.1 provides more details of the six classes of soil physical measurements appropriate to various soil use requirements. Square brackets are used in this table and elsewhere to indicate a functional relationship. Thus $\theta[\Psi_m]$ indicates that volumetric water content θ varies in some way with matric potential Ψ_m and, $\theta[-5\,\text{kPa}]$ indicates the water content of a sample that is at a matric potential of $-5\,\text{kPa}$. Methods of measuring most of the physical properties that are discussed in this section have been summarized and reviewed in Black (1965), Klute (1985) and Smith & Mullins (1990). Some of these methods are also described in British Standard 1377 (1975 – currently undergoing revision) and a greater range of standard methods of tests for soil properties is shortly to be produced by the 'soil quality' working group of the International Standards Organization.

Table 6.2 Soil physical properties of relevance to various uses

Application	Requirements
Playing fields	A, B & C
Footpaths	B & D
Gardens	C
Restored land for amenity use	C & D
Effluent disposal areas	A & D

Requirement	Relevant physical properties
A Drainage	Local hydrology, saturated hydraulic conductivity (and hence soil structure)
B Load-bearing capacity	Penetrometer resistance, bulk density, compactability, water content/potential when just 'playable', drainage and number of 'playable days'
C Use as a medium for plant growth	Drainage status, air capacity, available water capacity, bulk density, structure and/or penetration resistance
D Freedom from erosion and runoff	Infiltration, drainage, structural stability, (type of vegetation cover)

Physical properties and soil use

Table 6.2 provides some idea of the sets of soil physical properties relevant to soil use in urban areas; the list is not exhaustive. Many of the soil physical properties are interrelated in some way. For example, the saturated conductivity, k, decreases and strength properties increase, with any decrease in porosity (or increase in bulk density) that may be caused by soil compaction. The form of these relationships is itself dependent on particle size distribution.

It can often be difficult to obtain a quick, reliable measurement of a desired physical property (e.g. saturated conductivity) or to monitor changes in such a property with time. In these circumstances, measurement of a property that is more simply determined may provide an alternative. For example, bulk density is one of the easier physical measurements to make and may suffice; especially if it is possible to set an empirical upper limit on the value of bulk density which is acceptable for a given soil type and land use.

Particle size distribution

CLASSIFICATION

Some systems of particle classification which have been or are in common use are shown in Fig. 6.2. These systems represent a convenient way of referring to particles in any given size range. The upper limits of 'clay' and 'sand' size particles, of 2 µm and 2 mm are common to most systems of classification. Unfortunately, however, the size ranges referred to in the silt and sand fractions vary between systems, and the naming of the >2 mm size fractions also varies. This means that, where such names are used to identify particular particle size ranges, the classification system used must be specified to avoid ambiguity.

The actual procedure for particle size analysis of soils (e.g. Avery & Bascomb, 1974; BS1377, 1975) is long and tedious, and requires a skilled operator to produce reliable results. The results can also depend on particular aspects of the initial procedure used to destroy materials that are binding primary particles together. For example, where organic matter is not destroyed, some clay particles may be bound together in 'silt'-sized aggregates so that the proportion of clay is underestimated. In soils containing chalk or limestone, the normal treatment with 2 M HCl will dissolve this material and lead to an erroneous result. Even the selection of a small subsample of soil from a much larger bulk sample collected from a site, can limit the accuracy of the results. Mullins & Hutchinson (1982) have shown that, with the best available subsampling methods (e.g. chute splitting) but without replication, the accuracy to which the proportion in each size fraction is determined is typically ±10%. Thus particle size analysis results need to be treated with caution if it is intended to give particular significance to their precise values.

Particle size distributions may be expressed in terms of a 'particle size class' as shown in Fig. 6.3. This provides a convenient way of classifying and referring to the size distributions of soils but such names are inferior in terms of their information content to the original numerical results. Furthermore, because different national soil surveys use slightly different triangular diagrams it is possible for the same soil to be called by different names in different countries.

Particle size class is closely related to the 'texture' of a soil and, since the same set of names is used to describe each, it is often unclear which of these two properties is being referred to. Soil texture refers to the feel of a moist soil when rubbed between the fingers. Since sand, silt and clay each have a distinctive feel, an experienced person can make a remarkably good, subjective estimate of the particle size class of a soil from its texture. However, this procedure requires that other factors which can also influence the texture of a soil (e.g. organic matter and particular clay minerals) are taken into consideration (Hodgson, 1974). When this was done and surveyors calibrated their estimates of texture against the measured particle size classes of a set of soils from the study area, they were then able to obtain suprisingly good estimates of the percentages of silt and clay in a set of unknown soils (Hodgson *et al.*, 1976). For 80% of the soils, clay contents were estimated to within ±8% and silt contents to within ±12% of measured values. It can be concluded that an estimate of the particle size class of a soil by an experienced soil surveyor, working in his or her own area, can be preferable to the results from an indifferently performed particle size analy-

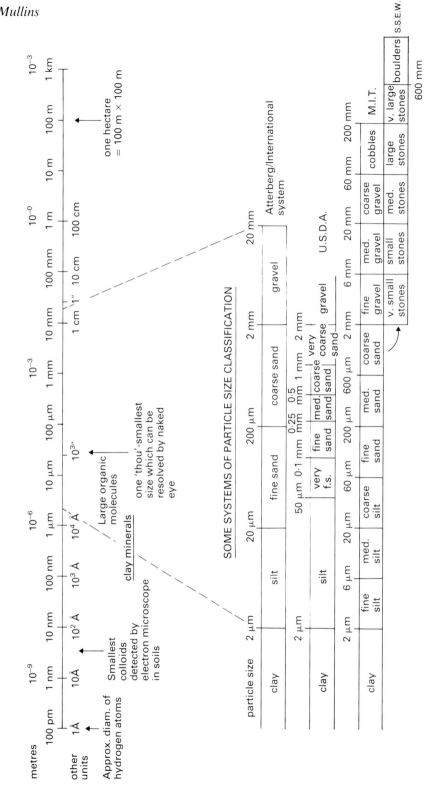

Fig. 6.2 Range of sizes commonly encountered in soil science (logarithmic scale).

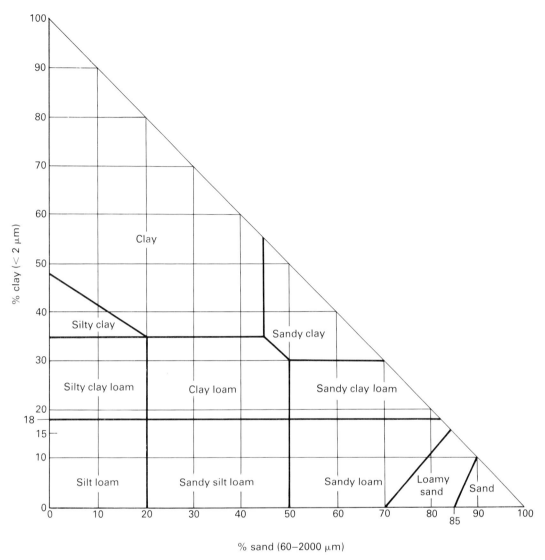

Fig. 6.3 Limiting percentages of sand and clay fractions for the particle size classes of the Soil Survey of England and Wales (Hodgson, 1974). Adapted by A.B. McBratney from the equilateral triangular representation to give a more easily used diagram.

sis; and the ability to 'hand texture' a soil reliably is something well worth learning and maintaining.

RELEVANCE TO OTHER PHYSICAL PROPERTIES

Particle size distribution has an indirect influence on most soil physical properties. In the case of transport properties, it is not the solid particles but the spaces between them that are important and particle size distribution is important only in so far as it affects pore size distribution and pore connectivity. In the case of soil strength, the existence of chemical and physical bonding forces between particles is of major importance together with the state of packing (i.e. porosity) of the soil and again, the effects of particle size distribution are only indirectly expressed. The relevance of particle size distribution is further complicated by the tendency of soils to be or become organized into structural units. For example, a compact, massive or poorly structured soil will have vastly different physical properties (both in terms of strength and water- and air-transport properties) than the same soil when cultivated to produce a seedbed made up of a loose collection of aggregates. Whilst this point may appear obvious, it must be remembered that: first, soil structural cracks can often extend to depths of 1 m and more in clayey soils (Reeve *et al.*, 1980); second, effective drainage of clayey soil (with or without the presence of field drains) is dependent on the maintenance of soil structure; third, damage to soil structure can build up progressively not only at the surface but also at depth where it is not directly visible; and finally, soil structural damage is often not easily identified before its effects become apparent from the inability of most roots to penetrate through the damaged layer or, from the build-up of a perched water table which holds up water in the surface layers of the soil and may lead to ponding and runoff.

There is thus a category of 'clayey soils' (which includes all particle size classes that have the word 'clay' in their name) in which structure exerts a dominant influence on many physical processes (e.g. infiltration, saturated water flow and gaseous diffusion).

At another extreme, sands, and some loamy sands and sandy loams, are structureless. With the exception of features such as worm channels or man-made holes produced by spiking, such soils are homogeneous. Their transport properties depend on the size distribution of pores and, unless there is interparticle bonding, their shear strength and compressibility depend on their porosity.

A further category of 'silty soils' is probably least well understood since the behaviour of these soils can be influenced both by structure and by their natural state of packing. Because silty soils shrink and swell little if at all in comparison with clays, any fine structural cracks which exist within such soils are vulnerable because, when damaged, they are less able to regenerate.

It is thus common practice to make a broad distinction between clayey soils, single-grain structureless (sandy) soils, and silty soils before attempting to relate particle size distribution to other properties. In the literature of soil mechanics, the first two of these are referred to as 'cohesive' and 'granular' materials respectively.

(a) Clayey soils

Because structural cracks in clayey soils are usually the dominant pathways for the surface infiltration of water, drainage, soil aeration and root growth, those features that determine shrinkage and the spacing between cracks are of considerable importance. Hodgson (1974) provided a useful classification system for describing the size of structural cracks (fissures), visible (macro) pores ($>60 \mu m$), and the distribution, diameter and type of roots, as observed at the face of a soil profile pit. When this system is used in conjunction with a description of the size and shape of peds (Hodgson, 1974), it provides a valuable record of soil physical properties at the time of observation that may be used for comparison with other parts of a site, other sites, or with profile conditions at a later date. This type of description can also be used in conjunction with soil texture to estimate saturated hydraulic conductivity (see later).

Shrinkage and swelling. Even in soil that appears to be dry, the spaces between clay particles can be water-filled. Soil shrinkage is a result of the increasing tension in the soil water which exists between the clay particles until the soil is almost air-dry. Most clay-sized (i.e. $<2 \mu m$) particles in soils consist of plate-shaped aluminosilicate clay minerals. Swelling in clays is thought to depend on the existence of an ionic double layer between the negatively charged surfaces of these clay particles and the balancing excess of positive ions in the soil solution immediately surrounding the particles. The net effect of the double layer is to cause a long range force of repulsion to exist between adjacent clay surfaces (a detailed discussion is given by Marshall & Holmes, 1988). The balance between this swelling pressure and the soil water tension which is pulling particles together determines the state of swelling in the soil.

Depending on the valency and concentration of cations in the soil solution, the double layer repulsion can extend from about 1 nm in concentrated solutions (e.g. in saline soils), to many tens of nanometers in the dilute solutions more typical of most soils. When it is realized that clay minerals are made up of repeated aluminosilicate layers about 1 nm in thickness, and that smectite particles (the finest clays) can often exist as particles with a few or even just one such layer, it can be readily understood how separating such particles by say 10 nm from one another could double the volume of a wet clay. In practice, this degree of swelling is not observed in soils because other forces, such as attractive chemical bonds and the overburden pressure (caused by the weight of the soil above) set a limit to the maximum amount of swelling *in situ*. However, where the chemical bonds are frequently disrupted by repeated disturbance of the soil (remoulding) in the presence of readily available water (i.e. in wet soil), swelling may continue far beyond its normal extent and the soil may be reduced to a sloppy almost liquid consistency (Mullins & Fraser, 1980). This is a part of the process of poaching (see later section) in which soils are damaged by repeated treading or wheeling when in a wet state.

From the explanation of swelling already given, it might be expected that the amount of swelling and shrinkage depends on the clay content and on the size of the clay particles themselves, and to a limited extent this is so. The size of clay particles is not usually measured but is inferred from the clay mineralogy. Thus, smectites which are the smallest of the common clay minerals (about 100 nm across and a few nanometres thick) shrink and swell the most. In contrast, kaolinites which shrink and swell much less are usually 0.5–2 μm across and 100 nm or so in thickness. In practice however, the situation is more complicated because clayey soils in the UK contain a range of clay minerals and the differences in mineralogy between many soils are not very marked (Avery & Bullock, 1977).

In a study of the amount of shrinkage which occurred in peds from clayey British soils under field conditions, Reeve *et al.* (1980) found that about 85% of the variation in shrinkage (between field capacity (−5 kPa matric potential) and air dryness) of peds from 19 topsoils and 42 subsoils, could be explained in terms of their dry bulk density at field capacity. Soils that had a higher bulk density showed less shrinkage. They suggested that the inverse correlation with bulk density is in part a reflection of the ability of certain clay minerals (notably the smaller mica-smectite minerals) to form more open structures.

However, from the point of view of the soil structural development that occurs as a result of the extraction of water by roots, it is only the shrinkage that occurs between field capacity and the wilting point (−1500 kPa matric potential) that is important since the soil is unlikely ever to dry much beyond this point except close to the surface (the upper 0.1 m). Here, Reeve *et al.* (1980) found little if any correlation between the shrinkage between these two limits and any of a whole range of soil properties (e.g. bulk density, clay mineralogy, percentage clay) that might have been expected to be relevant. Their conclusion was that, in practice, many factors influence the extent of shrinkage in the field, amongst which, the intrinsic soil properties are probably of minor importance. This seems to be borne out by field experience where the presence of a high water table (limiting soil drying and shrinkage), or the length of time that plant roots have to extend downwards may be of greater importance. Thus in a dry summer, tree roots which have had many years to grow down into the subsoil may extract water and cause structural cracking to the greatest depth, perennial grass may also crack soil to some depth whereas land which has been kept fallow and then sown to plants with a limited growing period or incomplete ground cover may only crack to a shallow depth.

(b) *Sandy soils*

In contrast to clayey soils, the physical properties of sandy soils are directly influenced by their particle size distribution and, even if soil structure is present, it is likely to be weakly developed and play a much less important role in soil aeration and drainage. Since pure sands or gravels are used as the fill for drain slots, in the construction of some sports fields, and in various sewage systems, their physical properties are of particular interest.

Bulk density. Dry bulk density is important because of its influence on other physical properties. For pure sand, bulk density is not dependent upon particle size as such but on the grading of the sand (i.e.

the broadness or narrowness of the particle size distribution) as well as on the shape of the sand grains. Narrowly-graded sands (such as those found in the Brecklands) are not very packable, especially if they are irregularly shaped. However, since there is a general tendency for smaller-sized sand grains to be more irregularly shaped this leads to an apparent relation between packing and sand size (Panayiotopoulos & Mullins, 1985). Broadly-graded sand or sand and gravel pack more closely because of the ability of the smaller particles to fit into the spaces between larger particles. This effect has been modelled by Gupta & Larson (1979). In sandy soils organic matter can play a major role in restricting the tightness with which the sand particles can pack together because plant remains and other organic debris restrict the ease with which the particles would otherwise pack. Furthermore, once there is even a small proportion of silt- and clay-sized particles present (i.e. loamy sands, sandy loams, etc.) packing and other physical properties are radically influenced because the forces of interparticle attraction become more important than particle weight (Smalley, 1964). Nevertheless it remains true that, whilst broadly-graded sandy soils (e.g. fine sandy loams) can possess excellent physical properties as a medium for plant growth, they are also much more vulnerable to compaction than narrowly-graded sands.

(c) Silty soils

The physical properties of silty soils are probably less well understood than those of sandy or clayey soils. Silts formed from platy micaceous minerals can behave quite differently to silts made up of less elongated particles. Although silts in East Anglia form some of the most productive of UK soils with good physical properties, it is also found that silty material in mine spoil heaps can often display many soil physical problems and be difficult to revegetate. The physical problems of such material are related to the low structural stability of many silts in the absence of organic matter. Thus, even if cultivated, the soil can easily slump or be compacted into a structureless medium that is poorly drained and is either too wet to allow adequate root aeration or too impenetrable when dry.

(d) Stones

Material with a diameter >2 mm is variously classified as gravel or stones (Fig. 6.2). On conventional soil survey maps it is easy to overlook the presence of such material unless it is so prevalent as to merit a particular mention in the legend. Since particle size analysis is conventionally performed on soil which has passed through a 2 mm sieve, the results of most analyses do not contain any reference to the presence of stones and it is only by consulting the profile descriptions or other more detailed accompanying text that any information on stone content can be obtained.

However, stone content may be easily estimated from the face of a soil profile or by sieving large soil samples in the field and weighing the stones (Hodgson, 1974). While it is conventional to treat the gravel/stone size fraction of soil as physically and chemically inert, this assumption cannot usually be made for soils in urban areas. For example, the rubble left after demolition of buildings will contain a mixture of fragments consisting of many of the following materials: brick, building stone, mortar, cement, concrete, plaster. This mixture may possess an appreciable available water capacity, cation exchange capacity, and may exert a considerable influence on soil physical and chemical properties. This behaviour is a reflection of the porous nature of some of these materials which have a much greater surface area (and hence cation exchange capacity) than non-porous stones of comparable size. Thus it is often inappropriate to treat stones merely as material that reduces the overall soil volume accessible to plant roots, and some consideration must be given to the physical and chemical properties of this size fraction. It may be difficult to obtain a reliable estimate of the available water capacity of such a stony soil, although Hollis (1987) does give tentative estimates for some types of porous and non-porous stones and rocks.

Bulk density values are of limited use as an indication of the likely ease of rooting in stony soil, even when a correction is made for the bulk density of the stones (Hodgson, 1974). Vine *et al.* (1981) showed that stones could increase the impedance to roots of a sandy soil so that stony/gravelly layers within the soil profile may exert an important influence by restricting rooting depth. Finally, it is worth remembering that stones may need to be removed from land that is to be used for playing grounds.

The particle size distribution of soil is related to its chemical properties, particularly cation exchange capacity, CEC. The reason for this is that CEC depends on the surface area per unit mass of soil which is itself dependent on clay content and type (Farrar & Coleman, 1967). Organic matter can also provide a significant component of the CEC of sandy soils. However, where there is likely to be a variety of non-soil based materials present, with CECs much greater or less than would be expected from their particle size, texture can give a misleading indication of the likely magnitude of the CEC.

Compaction

Like particle size distribution, bulk density is an important physical property because so many other soil physical properties vary with bulk density, particularly water and air transport properties and soil strength (Table 6.1).

BULK DENSITY AND POROSITY

Bulk density is a measure of how closely the soil particles are packed together but porosity (the volume of voids per unit volume of soil) is the most reliable measure of this property. However, porosities are less commonly reported than bulk densities because porosity requires an estimate or measurement of density of solids in addition to a measurement of bulk density (see Table 6.1). Furthermore, for the majority of soils that do not contain >4% organic matter, the density of solids lies within a narrow range. Thus bulk density can usually be closely related to porosity.

The bulk density of any horizon or layer within a soil is a variable that depends on soil management. Increasing soil bulk density increases its load bearing capacity but in most cases it also decreases the soil's ability to support and sustain plant growth. Thus there are many situations (e.g. playing fields) where a value of bulk density is attained that is a compromise between the conflicting requirements of load bearing (playability) and the necessity to sustain plant (grass) growth.

WHAT IS COMPACTION?

Unfortunately, confusion has been created by imprecise use of the words compaction, compact and compacted by some authors. Soil compaction is simply the process of reduction of the specific volume (or porosity) of a soil. Although this process often occurs as a result of surface loading such as treading, wheeling or raindrop impact, it is important to realize that soil which has been loosened may also undergo compaction as a result of settling and/or slumping. Settling can occur as a result of cycles of drying and wetting (and hence shrinkage and swelling) that allow aggregates to pack progressively more closely together (e.g. Mackie-Dawson *et al.*, 1989). Slumping occurs when soil aggregates weaken as a result of wetting so that partial disintegration at their points of contact allows them to pack more closely.

Acting in opposition to these compacting factors are the natural processes of frost heave and the effects of soil fauna as well as any man-made loosening. Frost heave is mainly of importance in the surface layers (i.e. 0–0.1 m) because of the limited depth of freezing under UK climatic conditions. A single night of heavy freezing has been observed to result in a 20 mm rise in the soil surface of a UK clay soil (Payne, pers. comm.). The major faunal agents in loosening are earthworms which bring up soil from depth (down to about 0.5 m) and void it on the surface in their casts.

The bulk density at any point in the soil profile is thus the outcome of these competing sets of influences, and references to compact, compacted or over-compacted soils are often no more than relative statements that the authors consider a soil more compact than some hypothetical natural or ideal state. On the other hand, it is quite reasonable to refer to a compacted soil where the soil has been subject to an applied load sufficient to cause compaction at some point in the profile.

It is important that roots are able to penetrate and extract water from the subsoil since the topsoil alone cannot provide sufficient water fully to satisfy plant requirements in lowland parts of the UK.

Thus, particularly in drier years and parts of the UK, plants that are not watered and that cannot root below 0.3 m are severely checked in their growth or do not survive the summer.

Because the influence of both compacting and loosening factors is dominantly within the top 0.3 m or less, subsoil bulk density changes only slowly if at all, except for cyclical changes caused by shrinkage and swelling in clayey subsoils. If roots and worms are able to enter a subsoil whose structure has been damaged, structural regeneration may occur due to the shrinkage caused by root water extraction and due to new worm channels. However, subsoils that are replaced in a very dense state or are severely compacted may be virtually impenetrable to roots and worms. These subsoils are unlikely to undergo structural recovery unless they are subsoiled or subject to some other form of deep loosening.

For example, Short *et al*. (1986) measured bulk densities of 1.4–2.3 (mean 1.74) t m^{-3} at 0.3 m depth for soils of the Mall (open parkland in the centre of Washington DC), and Craul (1985) found values of 1.52–1.96 t m^{-3} for subsoils in Central Park, New York. These are large values in comparison to most grassland areas but comparable values may well be found in many urban parks although, as discussed by both sets of authors, it may not be easy to decide between the various factors (such as the conditions under which extra soil layers are added and the subsequent human and occasional vehicle traffic) that may have resulted in these high bulk densities.

The topic of soil compaction has received considerable attention in the agricultural literature because of the serious limitations that compaction can impose on the establishment and growth of plants. These limitations are mainly the result of restricted soil aeration and of increased soil strength both of which can slow down or halt root growth. Soane (1982) provides an extensive review of the compaction by agricultural vehicles and the monograph by Barnes *et al*. (1971) on compaction of agricultural soils also includes an extensive review by Trouse on soil conditions as they affect plant establishment, root development and crop yield.

With the possible exception of narrowly-graded sands and gravels, all soils are vulnerable to compaction. In practice, when a load is applied to a soil

profile, it compacts to the point at which its strength is just sufficient to bear that load, and subsequent application of the same load under the same conditions will not produce any further increase in bulk density. However, since soil compressibility generally increases with an increase in water content (Larson *et al*., 1980), reloading a moist layer, previously compacted under dry conditions, will cause further compaction and increased bulk density. Thus for any given type of loading (e.g. treading), the important question is: how wet can the soil become before loading will increase the bulk density at any depth beyond some critical value?

Although it is not possible to be categorical about what is a critical value for bulk density, in many soil types there are very serious limitations imposed on root growth when bulk density exceeds 1.6 t m^{-3} (i.e. porosity is less than about 0.4), especially where the effects of the compaction have been to close up or block off cracks and other macropores. Apart from sandy and coarse loamy soils, soils with a bulk density >1.6 t m^{-3} have been found to have air capacities <10% (Reeve *et al*., 1973). Air capacity (Table 6.1) is a measure of macroporosity but, although values <10% have been considered to limit root aeration, the situation is complicated in structured soils where cracks that drain readily may be the main channels for soil aeration and root growth and such critical values may be misleading (see later). Thus, because of the importance of macropores as pathways for drainage, aeration and root growth, bulk density results should not be considered in isolation from the information gained from a profile description. In practice however, urban subsoils that have high bulk densities are likely to have been affected by types of disturbance and compaction which have also closed up most of their cracks and macropores. In these soils, where roots can penetrate at all, it is only down old root channels, occasional cracks or in weaker zones created by the decomposition of anthropic materials (Craul, 1985).

A model has been produced by Smith (1987) from which it is possible to predict the effect of any given load on the bulk density profile beneath the load. The current model is unreliable near the soil surface but has provided accurate predictions below 0.2 m. Whilst the practical applications of such a model are limited, its use on a limited number of soils of known strength properties (compressibility, etc.) is valuable because it allows a number of hypothetical compaction scenarios to be compared.

The effect of compaction on soil structure at depth is poorly understood. Blackwell (1988) has shown that cylindrical, vertical pores (that can be produced by coring or earthworms) and narrow, triangular or parallel-sided slots (and by implication vertical structural cracks) can survive well in soil under large vertical loads. More work on the stability of vertical pores and slots in relation to their dimensions, soil physical conditions and the type of surface loading, and in relation to the use made of them by plant roots will be helpful in enabling the design and use of coring and slotting machines to be optimized. Such techniques are of great importance because they hold out a possibility for soil amelioration in conditions that would otherwise prove intractable.

LOAD BEARING CAPACITY

Typical static ground pressures exerted by a human, tractor and a cow are 30, 100 and 200 kPa respectively. Since our feet are not always on the ground simultaneously and since, when we are running the ground also has to be accommodate a deceleration, it is reasonable to require a similar load bearing capacity for humans as for tractors and other off-road vehicles. However, whereas considerable deformation (e.g. 0.05–0.1 m) of the soil surface may be acceptable in an agricultural context, this would be quite unacceptable on a playing field and thus much greater attention has to be focused on the physical properties of the upper layers of the soil (0–0.15 m) both in terms of load bearing and in terms of soil damage. Estimating bearing capacity from penetrometer resistance, Schothorst (1963) found a value >700 kPa to be necessary to avoid pasture damage by cattle, whereas van Wijk & Beauving (1975) have assigned minimum acceptable values of 1000 and 1400 kPa respectively, for moderately- and intensively-played parts of a football pitch.

Large vehicle loads are carried by spreading the load over a large area with large tyres, double tyres or tracks. It is worth noting that, although this can result in similar values of ground pressure (<100 kPa) to that under smaller vehicles and humans, the effect on the soil may be radically different. This is because the lines of equal stress under a larger loading area extend to a depth in proportion to the size of the loaded area. Consequently, in contrast to the shallow compaction damage caused by humans and light vehicles, heavy loads such as a laden trailer with a 30 kN (3 t) wheel load can cause considerable compaction at depths of 0.3 m and more, as demonstrated by Smith (1987). This illustrates the potential damage that may be done by improper use of heavy earth moving machinery during land restoration.

POACHING

Poaching is an extreme form of compaction damage. It has been defined as 'the penetration of the soil surface by the hooves of grazing animals causing damage to the sward' (Patto *et al.*, 1978). The ultimate result is a trodden, uneven and hummocky surface of characteristic appearance in grassland. Although poaching is not relevant to sportsfields, this type of effect is apparent in narrow pathways and entrances, etc. wherever people or machinery are forced to tread repeatedly or track the soil at a water content close to field capacity. Scholefield & Hall (1986) have designed a recording penetrometer to simulate the action of the hoof of a walking cow and measure the resulting deformation of the soil. The apparatus gave results (Scholefield & Hall, 1985) which support the hypothesis that poaching results from a progressive loss of soil strength during repeated treading in wet weather. Interestingly, the severity of poaching was found not to be strongly related to the clay content of the soil but it was influenced by bulk density and the strength of the sward. The implications of this work are that simple measures to remove free water may be sufficient to reduce poaching damage greatly in areas which are trodden under wet conditions, although where treading occurs during rainfall or very shortly after some damage may be unavoidable.

Drainage

Theory and methods of drainage are dealt with in detail in Luthin (1957) and van Shilfgaarde (1974), and a more applied approach is given in Castle *et al.* (1984) and Farr & Henderson (1986). The purpose of drainage is to lower the water table after heavy rain to an acceptable depth in an acceptable time. For agricultural purposes this depth is commonly 0.5 m and an acceptable time may be one or two days.

Drainage serves at least three purposes. First, if the water table is allowed to persist for much longer periods within the bulk of the plant root zone, anaerobic conditions are likely to develop causing a delay in root growth and uptake of some nutrients, and ultimately root death. Second, the sooner the soil has been drained, the quicker it can be further dried to a point at which it has an acceptable load-bearing capacity. Finally, effective drainage provides a protection against runoff. It does this by maintaining sufficient drainable porosity in the surface layers to accommodate an intense rainfall event. Thus, for example, a drained soil with an air porosity of 10% to a depth of 0.5 m would be able to absorb a 50 mm rainfall event without runoff, providing the maximum infiltration rate of the soil surface was not exceeded. In practice, agricultural drainage systems are often designed to drain at a maximum rate of 10 mm day^{-1} so that such a rainfall event would take some time to drain away and, for more intensive land use such as playing fields, a much greater rate (e.g. 35 mm day^{-1}, Adams, 1986) is more appropriate.

Drainage problems can arise for a number of reasons and it is important that the source of the problem is correctly identified because this will determine the design of the drainage system. Diagnosis will usually involve a site inspection after heavy rainfall including the digging of inspection pits and, if necessary, the use of piezometers. The following types of drainage problem can occur: runoff from neighbouring higher ground or from broken drains; artesian water; low infiltration rate of the soil surface or of some compacted layer near to the surface (leading to a perched watertable); high local watertable; and low subsoil hydraulic conductivity.

Problems of runoff and broken drains are likely to be evident when a site is inspected after heavy rainfall. Cutoff (interceptor) drains or ditches can be used to divert the water but drains must be connected with gravel fill to the surface or to the layer causing the problem. Artesian water can be a difficult problem to diagnose as water may seep out only slowly and can appear apparently at random on a site and not necessarily at the lowest point. The existence of artesian water can be verified by the use of piezometers but, unless the aquifer responsible can be identified, the only solution is to drain each wet spot separately. Where a previously compacted layer can be identified, it may be possible to release a perched water table by subsoiling. Poor surface infiltration can also be improved by spiking or by providing an alternative flow pathway such as sand-filled slit drains. However, treatment of any or all of the first three problems will only be successful if there is not a groundwater problem or a subsoil of low hydraulic conductivity. Except on some sandy soils with deep profiles, good drainage is only likely to be achieved by a systematic underdrainage system (of clay or slotted plastic pipes) combined, if necessary in clayey soils, with a secondary treatment such as mole drainage.

Agricultural field drainage systems have too often been ineffective, partly as a result of the unsuitable timing or quality of the installation. Ministry advice on workmanship and materials (Ministry of Agriculture Fisheries and Food, 1983) provides a valuable set of guidelines. Apart from unsatisfactory conditions at the soil surface, drains that still flow at a substantial rate more than two days after rainfall provide an indication that the drainage system may not be functioning effectively.

An important distinction between an infiltration/surface drainage problem and that of lowering the local groundwater table is that infiltration is determined by the vertical component of saturated conductivity K_v, whereas the flow of water to drains only occurs below the water table and its rate is therefore determined by horizontal conductivity K_h. Infiltration is thus influenced by biopores (e.g. worm or previous root channels) and spiking or coring whereas drainage below the water table is not.

SATURATED CONDUCTIVITY, TEXTURE AND STRUCTURE

Despite the highly developed state of drainage theory, the drain spacing in most agricultural drainage systems is based on experience albeit codified into a set of empirical guidelines. The limiting factor has been the difficulty of obtaining reliable estimates for the saturated hydraulic conductivity K_{sat} of the subsoil through which the water has to flow to the drains. The difficulty in obtaining reliable estimates for K_{sat} in soil surface layers is also an obstacle to the successful diagnosis of problems caused by low infiltration/surface drainage rates. However, recent advances in our ability to estimate both K_v and K_h of soil horizons from profile examination hold out great promise for on-site problem diagnosis in the future and merit a brief discussion.

In the past, estimates of K_{sat} were often based on texture alone. McKeague & Topp (1986) demonstrated clearly the inadequacy of this approach by measuring *in situ* K_{sat} values between 0.1 and >10 m day^{-1} for both loamy sands and heavy clays, and for a range of soils with intermediate textures. Even when soil structure (pedality) was also taken into account, field estimates of K_{sat} were of little use because the descriptions of soil structure did not include any reference to the size of the cracks between the peds (McKeague & Topp, 1986) and this was often the most important factor controlling K_{sat}. However, where the horizon texture, structure, crack width, biopores, laminations and overall appearance (as an indication of compressed or cemented layers) were taken together, it was possible to develop successful systems to estimate both K_h and K_v (McKeague *et al.*, 1982; Wang *et al.*, 1985). These systems rely purely on field observations of the soil profile and have been found to estimate successfully the actual measured values of K_h and K_v. In over 80% of the soil horizons studied by these authors, hydraulic conductivity estimates were in the same or the next class to the measured values (where each class covered hydraulic conductivities that varied by a factor of about three times). Since measured values of K_{sat} show a high degree of spatial variability and are very time consuming, a system based on profile observation holds out great promise for the future, although training on sites where reliable conductivity measurements have already been made should greatly improve its reliability. The successful use of their system by McKeague *et al.* (1982) on a range of cultivated and otherwise disturbed topsoils implies that this type of system can be used (probably with modifications) for disturbed soils in urban areas.

CLAYEY SOILS

The topic of drainage and aeration of clayey soils is complicated by the fact that there is a seasonal variation in the width of their structural cracks. In the autumn, cracks produced after the soil has been dried out to depth by plant roots may persist and serve a useful function well into the winter but are likely to have closed up by early spring. For example, Leeds-Harrison *et al.* (1986) found that up to 40 days of continuous saturation was required to allow a clay soil to swell fully. Without cracks, the saturated hydraulic conductivity of clays and most clayey subsoils is so low that even close-spaced (i.e. 5 m or less), pipe-drainage systems are ineffective. In practice, the only effective system (in an agricultural context) has been to use a mole drainage system in which a 50–75 mm diameter bullet-shaped object is drawn through the soil at a depth of 0.5 m or less and at a spacing of between 1 and 3 m. When carried out in a suitable soil under appropriate conditions, mole drainage can be very effective because the mole plough not only produces closely-spaced drainage channels that may be stable for five years and more, but also produces a system of artifical cracks above and radiating from the mole drain that channel water directly from the soil surface. The mole drains, which are drawn at right angles to intercept a series of widely-spaced (20–80 m) gravel-filled slots above conventional pipe drains, may be redrawn at regular intervals (e.g. five years) or whenever they are thought to have become ineffective.

Effective mole drainage is dependent on sufficient stability in the soil around the mole channel for it to remain open for a number of years and for the connecting system of cracks to remain in working order. Because failure of the system can occur in a number of ways, it does not seem likely that any simple soil test will be sufficient to indicate the suitability of a soil for mole drainage although structural stability tests can indicate when a soil is likely to be unsuitable (Spoor & Ford, 1987). However, as a rough generalization, the greater the content of clay, the more likely that mole drainage will be successful.

Where mole channels are unsuitable, shallow gravel-filled moles have been used and the even more intensive system of sand-filled slit drains used on sports grounds is a logical extension of such a system to the situation in which artificial cracks cannot be relied upon because they would soon be closed up by the action of treading.

Soil as a medium for plant growth

PLANT REQUIREMENTS

The soil physical requirements of plants have been discussed in a wide range of soil and plant science textbooks (e.g. see Marshall & Holmes, 1988, for a good summary or Wild, 1988, for more detail on plant functioning) and only a brief summary including some recent developments is given here. Plants require a suitable temperature range, support to stand upright but not too much resistance to shoot emergence or root growth, a readily available water supply to roots and adequate root aeration. In the special case of plants adapted to wetland conditions, the last requirement may be satisfied by oxygen transport within the plant rather than through the soil. Maps of accumulated temperature and average maximum potential soil moisture deficit are available for Scotland (Birse & Dry, 1970), and England and Wales (Jones & Thomasson, 1985). Birse (1971) has produced a bioclimatic map of Scotland that also includes exposure and frost risk.

Plants are particularly sensitive to soil physical conditions at germination and during the early stages of growth when they have a small root system growing in a shallow depth of soil. In particular, during a dry period the rate of root extension must exceed the rate at which the soil is dried out. It is therefore understandable that, when the soil is used for establishment of seedlings, greater emphasis is placed on its physical properties within the top 0.05–0.1 m depth than at greater depths. The requirements of seedlings are usually met by placing the seed or seedling in a loose layer of more or less aggregated soil, just sufficiently compacted to give adequate seed–soil contact and to support the emerging seedling.

Because drying of the soil profile from a bare soil surface proceeds only very slowly once the top 0.05–0.1 m of soil has dried out, it is not essential for plant roots to be able to grow at such a fast rate below a depth of 0.1 m. Both restricted aeration and mechanical impedance commonly slow down the rate of root growth under many soil conditions.

(a) Soil aeration

Oxygen flow to roots is dominated by the process of diffusion in the air-filled fraction of the soil pores. The rate of diffusion is proportional to the air (filled) porosity and inversely proportional to the tortuosity (a measure of pore interconnectedness). In the summer, when the rate of oxygen consumption by roots and microorganisms is high, an air porosity value of 10% has often been quoted as the value below which oxygen supply becomes limiting to root growth. However, much smaller values of air porosity are sufficient to maintain aeration to roots growing in the vertical cracks and biopores of a structured soil or growing at lower soil temperatures. Because reducing conditions brought on by anaerobism can result in soil chemical conditions likely to cause root death, a simple field test that can indicate the presence of ferrous iron and hence of reducing conditions (Batey & Childs, 1982) is of value as a quick field diagnostic test.

Providing plant roots are adequately oxygenated, it does not matter how wet the soil is. Thus submerged roots that are supplied with a continuously oxygenated solution, in a hydroponic growth system, or because they are growing in a 'flush' site on a slope where aerated water flows through the soil, can make excellent growth. In practice such situations are uncommon and the movement of soil water is rarely such as to be able to supply much of the root's oxygen requirements. A simple index of soil quality in terms of aeration is the air porosity of the soil at a matric potential of $-5 \, \text{kPa}$ since this is a typical value of the potential that wet topsoils are likely to drain to shortly after rainfall. Methods for determining this air porosity (sometimes referred to as the *air capacity*) and average values for a wide range of topsoils and subsoils, grouped according to particle size class, are given in Hall *et al.* (1977). This shows that sandy soils rarely have air capacities <10% and are therefore unlikely to exhibit aeration problems unless there is a persistently high groundwater table.

(b) Mechanical impedance to roots

Because dry bulk density is a measure of the closeness of packing of soil solids, many researchers have sought an upper limit to the bulk density of a soil beyond which root penetration is not possible. In the much quoted work of Veihmeyer & Hendrickson (1948), nine moist soils of contrasting particle-size distribution were packed into metal cylinders at two different values of bulk density in order to determine the critical density above which sunflower roots would not enter the soil. No roots penetrated sandy soils with bulk density of $1.9\,t\,m^{-3}$ or more and there was no penetration into the clayey soils when the bulk density reached values of $1.6–1.7\,t\,m^{-3}$. These results have been taken by some authors to indicate that there is a smaller critical bulk density for clays than for sands. However, in practice roots can often be found growing down structural cracks in clayey soils with bulk densities of up to $1.8\,t\,m^{-3}$; Veihmeyer and Hendrickson used remoulded soil that did not crack.

The rate of root growth decreases steadily in soil of progressively greater mechanical impedance until their turgor pressure is insufficient to allow further growth (about 1 MPa). In comparatively structureless soil, penetrometer resistance provides a valuable guide to the impedance likely to be experienced by roots. Because a penetrometer is not a perfect root analogue and, in particular because of the greater frictional resistance of the penetrometer tip, it is found that penetrometers experience between two to eight times as much resistance as plant roots growing in the same soil (Whiteley *et al.*, 1981). Thus it is found that in soil with a penetrometer resistance of 3 MPa or more, root growth is halted or severely retarded (Greacen *et al.*, 1969).

Root growth in structured soil is poorly understood although it has been found that roots are able to locate cracks (Dexter, 1986a) or cylindrical pores (Dexter, 1986b) at the boundary between a loosened layer and a compact subsoil, and grow down these pores. The important distinction between biopores (i.e. channels left by worms and decayed roots) and structural cracks is that biopores exist independent of the water content of the soil whereas cracks open up in the spring as a result of the extraction of water by plant roots. Indeed extraction of water by roots can generate structure in a previously structureless soil although the soil conditions that determine whether or not this will happen are not fully understood (Young *et al.*, 1988). The mechanical stability of biopores against swelling and shrinkage in comparison to the more seasonal variation in the width of cracks is a consequence of their shape, although their sides may also be stabilized with organic residues and therefore more water stable than artifically made pores.

(c) Water supply to roots

In an unsaturated soil the water is held under a negative pressure (tension) referred to as the matric potential. Plant roots have to exert a greater tension than this in order to extract water from the soil. The point at which a plant is damaged by water stress so that it does not recover when watered is known as the (permanent) wilting point. For experimental convenience this is operationally defined as the water content of a soil sample that has been equilibrated at a matric potential of -1.5 MPa although, because there is some dependence on soil and atmospheric conditions, this can only be an approximation. Even plants grown in soil at much higher matric potentials (e.g. -100 kPa) may exhibit partial wilting on hot dry days and grow at less than optimal rates if other factors are not limiting.

THE SOIL WATER REGIME

(a) Available water

The soil profile acts as a reservoir that plants rely on for a steady supply of water. As roots extract water from the soil the water remaining is held at progressively lower potentials and therefore becomes progressively less easily obtainable. As a consequence there is a physiological response in which a stressed plant partially closes its stomata and thereby reduces transpiration.

The size of the soil reservoir is called the profile available water (PAW, measured in millimetres) and is determined by the rooting depth and by the available water capacity of each layer of soil within the rooting zone. Roots may also extract water from a greater depth by unsaturated flow but this rarely provides more than a few per cent of their requirements except where there is a water table maintained throughout the summer within a short distance of the rooting depth.

The maximum plant rooting depth is often restricted by soil physical or chemical factors rather than the physiology of the plant although the distribution of plant roots with depth can vary markedly from species to species. Where mature plants are already established, it is possible to estimate their maximum rooting depth by a careful examination of the face of a soil profile pit. In the absence of plants, it is still possible to identify potential barriers to rooting such as dense/impenetrable layers or high or perched water tables. A penetrometer is useful in identifying a mechanically impeding layer especially in sandy soils or soils with a cemented horizon. Since sand may flow freely through the fingers but still be sufficiently closely packed in the profile to be impenetrable to roots, profile examination without a penetrometer can be misleading.

The available water capacity (AWC) of a soil sample is the water content that may be stored in that sample in a form available to plants. It is defined as the difference between the field capacity and the wilting point. The field capacity itself is defined as the water content at any point in a profile that has just ceased draining (usually taken as one or two days after heavy winter rainfall). Since there can be no precisely defined endpoint to drainage, the definition is imprecise but the concept is an extremely useful one. Even clay soils have been shown to drain regularly to a more or less constant profile water content during the winter (Reid & Parkinson, 1987; Mackie-Dawson *et al.*, 1989). Webster & Beckett (1972) showed that the matric potential to which many topsoils in south central England equilibrated after drainage was about $-5\,kPa$. Thus the water content of a soil sample that has been equilibrated at a potential of $-5\,kPa$ is taken as an operational definition of field capacity for soils in the UK.

Methods for measuring AWC are described in a Soil Survey monograph that also provides values from a very large number of samples from England and Wales and a table of average values for topsoils and subsoils in terms of the particle size class and packing density (Hall *et al.*, 1977). Unfortunately, the use of the term 'packing density', which is defined in terms of bulk density adjusted by an empirical term allowing for clay content, imposes a limitation on the ability to which a physical understanding may be obtained from the data presented in many Soil Survey publications because it combines two distinct soil physical properties. However, in general terms, soils with low, medium or high packing densities will have corresponding values of

dry bulk density with reference to an arbitrary set of norms, and Hodgson (1974) describes a simple procedure for estimating packing density in the field.

Hall *et al.* (1977) show that, for a wide range of soil textures, average AWC values range between 13 and 25% volumetric water content with larger values associated with silty soils and with samples with a smaller bulk density (generally topsoils). However sands, especially coarse sands, may have considerably smaller AWCs (as small as 5%) and occasionally less. Hollis (1987) has extended the data of Hall *et al.* and presents a more easily usable set of tables from which to estimate AWC from soil texture and structure (without reference to packing density). He also gives AWC values for humose and peaty soils.

Profile available water (PAW) is simply obtained by dividing the profile, down to the limit of rooting, into a number of layers for each of which the AWC is estimated or measured. The contribution of each of these layers is then calculated as the product of its AWC (expressed as a fraction on a volume basis) and its thickness, and all contributions are summed to give the PAW. In performing this calculation, the AWC values for deeper horizons are sometimes reduced by an empirical, plant-species-related factor, intended to account for the sparser rooting at depth which is considered to place an added limitation on the water available to the plant (e.g. Hollis, 1987).

(b) Plant water use

The combination of plant water loss and water loss through the soil surface is referred to as evapotranspiration. For short, green crops fully covering the ground, that are well supplied with water and in an open situation, the evapotranspiration is found to be determined only by the weather (radiant energy, relative humidity and wind speed) and is referred to as potential evapotranspiration (PE). This quantity can be deduced from meteorological data, and weekly values are commercially available for all parts of the UK from the Meteorological Office. Their computations are based on a model referred to as MORECS that also produces empirically adjusted values for actual evapotranspiration from other vegetation types (e.g. coniferous forests). The model is described in Thompson *et al.* (1981) and a discussion of some of its limitations and applications is given in Gardiner (1981). Approximate values for PE for different parts of the UK may also be obtained from Ministry of Agriculture Fisheries and Food (1967). Typical values for most of lowland Britain during the summer are 3–4 mm day^{-1}, and much lower in the winter.

(c) The soil water budget

By using the water content of the profile at field capacity as a reference state, it is possible to construct a soil water budget. During winter, when weekly rainfall exceeds evapotranspiration the soil remains at or near to field capacity but at some point in the spring, evapotranspiration exceeds rainfall and a *soil moisture deficit* (SMD) starts to accumulate. This deficit is the number of millimetres by which cumulative evapotranspiration exceeds the cumulative rainfall, after the date at which the profile first left field capacity.

In lowland Britain the soil returns again to field capacity usually during the autumn although in exceptional years (e.g. 1972/1973) a deficit may be carried over from one year to the next. Median dates at which the soil leaves and returns to field capacity (under a crop transpiring at the 'potential' rate) are available for different parts of England and Wales (Smith & Trafford, 1976). Actual dates for any given year and for a variety of vegetation types as well as weekly values of SMD, calculated from the MORECS model, are commercially available from the Meteorological Office. However, it is more accurate to calculate local values of SMD if reliable rainfall records are available locally since the spatial distribution of rainfall is variable and the rain gauging stations of the Meteorological Office are widely spaced in some parts of the UK. For this and other reasons, the date of return to field capacity can only be taken as a very rough estimate and it would be necessary to monitor soil conditions with a tensiometer if a more reliably estimated date were required.

Because water content also determines soil strength and hence walk/playability, the period in the year when the soil is at or wetter than field capacity (the number of field capacity days) has great significance for a wide variety of applications.

A map of the mean duration of field capacity in England and Wales (Jones & Thomasson, 1985), whose results should be representative of the grassland situation, has values for Ipswich and Colchester of <100 days per year whereas such diverse places as Swansea, Manchester and Plymouth are all >200 days per year and in marked contrast to the greater London area (typically *c.* 125 days per year).

The Soil Survey of England and Wales classify soils into one of six *wetness classes* (Hodgson, 1974) based on the average number of days per year that the soil is wet to within 0.7 or 0.4 m of the surface (where 'wet' refers to the presence of a water table detected by a dip-well or tensiometer). In many cases a good indication of the wetness class can be obtained from morphological features of the profile including gleying (shown by soil colours), horizons of low saturated hydraulic conductivity and clay content. However, there are complicating factors with such an approach such as the existence of an effective field drainage system, and, in red soils, gleying is more difficult to observe (Robson & Thomasson, 1977). Since wetness class is closely related to field capacity days (Robson & Thomasson, 1977), a trained soil surveyor may be able to estimate field capacity days simply from profile examination where the profile is not too disturbed or mixed with non-soil material, and a field methodology for estimating wetness class is currently in preparation by the Soil Survey (Hollis, pers. comm.).

The maximum SMD in any year represents the driest state reached by the profile. The maximum potential SMD is calculated using values of PE and, when averaged over a period of years, it is a valuable guide to the climatic demands placed on the soil profile by growing plants. For example, the average maximum potential SMD for lowland Aberdeenshire is about 65 mm whereas it is >200 mm for parts of East Anglia. This means that in most summers, a profile providing a rooting depth of 0.4–0.5 m is sufficient to supply the water requirements of plants on all but a few sandy soils in Aberdeenshire, whereas plants grown on the same profiles in East Anglia could exhibit pronounced stress symptoms or even die.

In parks, gardens and other enclosed areas, water budget calculations can only be regarded as a rough estimate of soil conditions and on-site measurement of matric potential (e.g. with tensiometers) and/or water content (e.g. with a neutron probe) are necessary to obtain more precise and reliable information.

Specialist topics: sports fields and amenity land

There is a continuum in soil use that stretches at one extreme from sports fields that are constructed to allow the maximum amount of use especially in winter (e.g. football pitches of the major clubs), to parks and other grassed areas normally reserved for amenity use and never or only occasionally used as playing fields. Thus economic factors and the intended timing, frequency and flexibility of use determine the importance that may be attached to the soil physical properties of 'sports fields' (here taken to include amenity grassland). However, when reclaimed or derelict land is being reinstated, it may be as economical to create a high quality sports field as to reinstate the site for any other amenity use (Adams, 1986). Even where it is not intended to create a sports field, for any reinstated or renovated area that will have to bear heavy use in wet conditions, the physical properties and drainage of the replaced 'soil' can now be optimized and wear-resistant grass species can be selected. Thus there is a wealth of literature on physical properties of sports fields that is worth considering not only for its own sake, but also because the findings may be adapted and applied to the more varied and usually less demanding requirements of urban parks and other grassed open spaces (e.g. see Adams, 1982).

The three factors that have to be brought into balance in considering the allowable intensity of use of any sports field consistent with sustainable use are: (i) playability (which includes load bearing capacity); (ii) drainage, and (iii) the requirements for sustained growth of grass. These factors have already been discussed in general terms previously but this section is concerned with the more specialist literature, particularly that concerned with removing soil physical constraints and allowing greater use of sports fields.

The past 20 years have seen rapid progress in the application of physical principles to the rational design of sports fields so that the major limitations to designed sports fields lie, not so much in an absence of scientific criteria and specifications on which to base a design, as in the difficulties of obtaining suitably sized and graded sands.

In the UK there is a specialist sports field research centre (the Sports Turf Research Institute, Bingley, West Yorkshire, BD16 1AU), originally established in 1929 to do research and provide an advisory service for golf clubs, and extended since 1951 to include all types of sports field applications. The institute provides a specialist advisory service on all types of sports and amenity turf (to their subscribers), and publishes a research journal. Useful practical advisory pamphlets and booklets are available from the National Turfgrass Council (3 Ferrands Park Way, Bingley, West Yorkshire, BD16 1HZ) on topics such as 'Sand construction for sports pitches', and from the Scottish Sports Council (1 St Colme Street, Edinburgh, EH3 6AA). British Standard 3969 (1965) recommendations for turf for general purposes are currently being revised.

DRAINAGE AND SOIL AMELIORATION

For a sports field to be playable it must be free from surface water. With the exception of some sands, infiltration and drainage of the surface soil layer requires the existence of soil aggregation or some structural cracks or pores to conduct water at a high rate. However, as Adams (1986) has explained, 'aggregate stability is insufficient to tolerate the disruptive action of play during wet weather. The degradation of structure in the immediate topsoil causes a marked fall in hydraulic conductivity leading to ponding. Soil strength diminishes markedly, play disrupts turf cover and the surface becomes plastic and slippery'. To avoid damage to the soil and the turf, sportsfields usually have a key requirement for a surface drainage system to ensure the rapid removal of excess water (i.e. free or loosely held water that would otherwise drain away too slowly). However, many of the pitches run by local authorities and clubs were created from native pastures and incorporate field drainage systems based on the agricultural design of pipe drains. They are not purpose-designed and suffer from surface damage under heavy or untimely use.

There is a range of commercially available implements whose aim is to improve the aeration and infiltration rate of the surface layers of soil, where most of the compaction caused by playing or walking occurs, including spikers, implements to remove small cores of soil, surface slotters and shallow subsoiling devices (see Fig. 6.4). These implements should reduce ponding, increase porosity in the surface layers and improve conditions for grass growth but their effects may be short-lived, especially if the field is used again under wet conditions.

For sports fields intended for more intensive use in the autumn, winter and spring (i.e. in conditions that are likely to be wet), there are several possible techniques to improve drainage and the playing quality of the ground (see Fig. 6.4) which have been reviewed by Ward (1983) and Adams (1986). These range from the installation of sand-slit drains on top of, or together with, a conventional pipe under-drainage system with gravel fill on top of the pipes; to the addition or incorporation of sand onto or into the surface of the pitch (sand construction). Finally, there are specially constructed systems (e.g. the Cell System, Purrwick and the Prescription Atlantic Turf System) consisting of a sealed plastic-lined depression filled with carefully selected sand or sand mix; these are also capable of subirrigation (see Adams, 1986 or Ward, 1983 for further details).

Sand-slit drains serve partly to increase the design specifications of a conventional pipe drainage system by effectively reducing drain spacing and partly to intercept surface water directly. If they are not installed so as to connect with a gravel-filled slot on top of pipe drains then they will only be effective where they make connection with an underlying permeable layer. They are economical in the use of sand and gravel but there is a possibility of deterioration in surface trueness in clayey soils. The drainage improvement conferred by the sand-slits is relatively short-lived because of surface sealing of the slit system as a result of the movement of adjacent soil during play in wet conditions (Adams, 1986). Recommended design specifications for sand-slit drains are given by both Ward (1983) and Adams (1986).

Two types of sand amelioration are possible. First, there is the addition of a layer (of between 100 and 150 mm) of carefully specified sand to the surface of soil in which a sand-slit drainage system has been installed, as in the Prunty Mulqueen patented system. Second, sand and soil can be mixed in proportions such that there is no more than 20% of particles $<100\,\mu m$. With a properly specified sand, the result is a surface layer combining a high hydraulic conductivity, reasonable available water capacity and air porosity, and adequate shear strength.

(a) Physical properties and behaviour of sandy layers

Baker (1983) studied the effect of the physical properties of sands on the properties of sand-soil mixtures of the type likely to be used for sports field construction by sand amelioration. Three soils, a sandy loam, a sandy clay loam and a clay loam

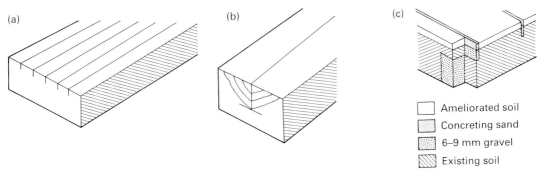

Fig. 6.4 Some of the techniques to improve infiltration, drainage and playing quality of sports fields: (a) surface slotting, (b) loosening/vibratory fissuring, (c) (redrawn after Adams, 1986) combined soil amelioration and sand/gravel slit scheme. Sand/gravel slits 0.3 m deep at 1 m centres connecting through to 25 mm diameter permeable backfill over existing drains. Surface 0.12 m of soil is sand ameliorated to give D_{20} of 100 μm.

were mixed with a wide variety of sands in proportions such that there was 20% by mass of material <125 µm in the overall mix. Each mixture was equilibrated at a matric potential of −4 kPa and subjected to a standard compaction treatment with a falling weight to simulate the state that it would be in after a period of use. Despite the constant proportion of material <125 µm, his results provide a dramatic illustration of the importance of the physical properties of the sand in determining the overall behaviour of the mix. Thus for the mix of sand and sandy clay loam, for example, saturated hydraulic conductivity varied from 5.8 to 156 mm h^{-1}, porosity varied from 0.31 to 0.39, air porosity (at −4 kPa potential) from 0 to 0.18, available water capacity (−4 kPa to −1.5 MPa) from 0.13 to 0.30, and vane shear strength from 9 to 27 kPa, depending on the properties of the sand. As might be expected, both hydraulic conductivity and air porosity increased, and shear strength and available water capacity decreased, with increasing size of the sand. Grading had a lesser effect, and increasing the proportion of angular grains increased the shear strength. Ward (1983) has reviewed the literature concerning the specifications of sand suitable for sand profile construction and amelioration and has concluded that, because of the narrowness of the specifications, a laboratory particle size analysis of representative samples is strongly advised. According to Adams *et al.* (1971) the sand should have less than 10% by mass of particles outside the range 0.1–0.6 mm and should be narrowly graded (such that the diameter of 90% of the sand grains should be not more than 2.5 times the maximum diameter of the smallest 10%). Other researchers have made similar recommendations (Ward, 1983).

Addition of sand to the soil surface can serve a useful purpose in delaying the sealing of sand-slit drainage systems (Adams, 1986). It can also improve playability once a sufficiently thick layer has been built up. However, on its own, on a flat site, a sandy surface layer on top of a less permeable soil cannot improve surface drainage unless it connects with a suitable drainage system.

The maximum soil water tension that can be caused at the surface, as a result of drainage, is proportional to the depth of the drains. This depth is the limit to which drains can lower the water table and, in practice, the water table may fall quickly after heavy rain to only some distance above drain depth. In this case the potential at field capacity is related to drain depth. However, the sand itself will not drain until a matric potential equal to its 'air-entry' value is reached. This potential is inversely proportional to the diameter of the pores between sand grains. Table 6.3 shows that, for example, although 0.1 mm diameter narrowly-graded sand has a hydraulic conductivity greater than the highest recommended value (of 10–100 mm h^{-1}), it is unsuitable as a surface layer because it requires a potential <−9 kPa to drain (Adams, 1986). This potential corresponds to a water-table depth of >0.9 m which is usually quite impracticable, and surface layers therefore require a low air-entry potential to drain quickly.

For more broadly graded sands or sandy soils, the fine material fills the spaces between the larger particles resulting in a much broader pore size distribution with no clearly defined air-entry potential and a substantial proportion of smaller pores that do not drain at high potentials (i.e. low tensions).

The air-entry potential for gravel is virtually zero. At potentials below this value the gravel has a negligible hydraulic conductivity and behaves as an impermeable layer. Thus, whilst (sand-covered) gravel slits that connect to the soil surface can quickly conduct away excess water, a continuous layer of gravel or coarse sand underlying soil at a shallow depth (e.g. <0.2 m) can have a disastrous effect on drainage if air-entry in the gravel occurs when the topsoil still has a high degree of saturation. At a slightly greater depth however, gravel underlying sand can create a perched water table in the lower part of the sand thereby increasing the profile available water and aiding grass growth.

PLAYING QUALITY

The qualities of the turf/surface soil combination which make it suitable for any particular sport are referred to as its playing quality or playability. Definitions of the physical properties involved are given in Table 6.4. These can be grouped into factors that influence interactions between a player and the surface, and factors that influence the interactions between a ball and the surface. According to the particular sport, one or other of these set of

Table 6.3 Some properties of compacted single-size rounded sands (from Adams, 1986)

Particle size (µm)	Saturated hydraulic conductivity (mm h^{-1})	Air-entry potential (kPa)
100	150	−9.0
200	450	−4.0
400	1 500	−2.2
800	5 500	−1.2

factors will take priority. Thus for sports involving running and sudden changes of direction (e.g. football, rugby) the first set of properties assumes priority, whilst for many other sports (e.g. golf, bowling, cricket) the manner in which the ball bounces or rolls is of primary importance (Bell *et al.*, 1985).

The various aspects of playing quality depend on the interaction between the turf and the soil surface so that it is difficult to separate the two except at a very general level. For example, grass roots can make a major contribution to the traction (or shearing resistance between a studded boot and the pitch) whereas the stiffness (or deformability) of a pitch will be greatly influenced by soil water content and the texture of the soil. There is a large body of empirical observations on playing quality but the current need is first to standardize tests for the various aspects of playing quality and then to relate these to the physical conditions of the soil (e.g. porosity, particle size class, matric potential) and of the grass sward (e.g. height, rooting density, surface wetness) in order that a more comprehensive overview can emerge. A detailed review of the techniques for measuring the playing characteristics listed in Table 6.4 is given by Bell *et al.* (1985). It is clear from this paper that the development and standardization of measurement techniques are well advanced to the point at which it is now possible to put acceptable limits on the values of many of these indices of playing quality for a variety of sports.

Table 6.4 Definitions of the various measures of playing quality (adapted from Bell *et al.*, 1985)

Property	Definition	Comment
I Player–surface interactions		
Friction	Surface friction to smooth soled footwear	High friction/traction can cause feet to become 'locked' to the surface causing ankle and knee twisting injuries. With low friction/traction players can easily slip over
Traction	Surface friction to footwear with studs, spikes or cleats	
Stiffness	$\dfrac{\text{Applied stress}}{\text{Deflection of the surface}}$	High stiffness causes muscle soreness, jarring of limbs and greater risk of injury in a fall. See Panayiotopoulos and Mullins (1985) for discussion of how to relate impact and static loading tests
Resilience	$\dfrac{\text{Kinetic energy after impact}}{\text{Kinetic energy before impact}}$	Low resilience leads to fatigue
II Ball–surface interactions		
Ball bounce	$\dfrac{\text{Height of rebound}}{\text{Drop height}}$	This depends on the properties of the ball and the surface. Cricket pitches with high resilience are dangerous
Rolling resistance	Operationally defined in terms of rolling distance of a ball released at a given velocity	
Friction and spin	A complex topic, see discussion in Bell *et al.* (1985).	

Where playability measurements are required in order to define or measure the suitability of a pitch for a given sport, tests that attempt to simulate the actions (of a ball or foot) involved in that sport must be of prime importance. However, where it is intended to relate soil physical measurements to soil and sward physical properties in order to understand why any particular pitch or part of a pitch has a given playability, measurements such as penetration and shear vane resistance which can be more easily related to soil physical properties may be equally important. Ultimately, the development of models on the basis of which playability can be predicted from soil and sward physical properties should provide the next advance in our understanding.

This type of approach has been taken by van Wijk & Beauving (1975, 1978) who have related penetrometer resistance (30° semi-angle, 1 cm² cone) to an empirical ranking scale for playability based on visual assessment of the effect of kicking the heel of a boot into the pitch. In this way they were able to assign minimum acceptable penetration resistances of 1.4 and 1.0 MPa respectively, for intensively and moderately played parts of a football pitch. Van Wijk and Beauving then proceed to show how penetration resistance varies with bulk density and matric potential. This seems to be a promising approach for two reasons. First, matric potential can be easily monitored with tensiometers in contrast to the time-consuming nature of penetrometer and other playability measurements and can also be related to the adequacy or otherwise of the performance of the drainage system. Second, it is possible to study the relation between penetration resistance, bulk density, and the particle size distribution and organic matter content of the soil both empirically (van Wijk & Beauving, 1978) and possibly also with a simplified theoretical approach (Mandiringana, 1984). At present, both of these approaches have only been applied to soil behaviour under wet (i.e. $\Psi_m > -10$ kPa) conditions but penetration is not a problem in drier conditions although hardness may be.

Adams *et al.* (1985) have studied factors influencing the stability of sports turf root zones. Apart from the differences in root production between different turfgrass cultivars they also found that roots increased the surface shearing resistance (traction)

of sandy root zones by a factor of two to three times ($\Psi_m = -3$ kPa) which was a much greater effect than could be achieved by increasing the silt and clay content of the soil within the acceptable limits.

WEAR

It is difficult to distinguish between the direct effects of wear on sportsfields and the indirect effects, caused by increased surface compaction with the consequent increases in bulk density, reduction in infiltration rate and air porosity, that decrease soil suitability as a medium for plant growth. Canaway (1976, 1982) has developed a differential slip wear machine capable of realistically simulating the wear caused by studded, spiked or smooth-soled footwear. This machine consists of a weighted trolley mounted on four wheels, the front two of which are driven at a different rate in a fixed ratio to the rotation rate of the rear wheels thus simulating a shearing action. Studs, spikes or cylindrical rubber soles may be attached to the wheels to simulate different kinds of footwear. The damage to grass by footwear without studs or spikes probably results from shearing which causes abrasion of the grass whereas studded footwear may also tear plant roots.

In walking, a minimum of horizontal (shear) force is applied to the ground whereas in running the component of shear is increased (Cavagna *et al.*, 1964) and, of necessity shear forces are applied when running and changing direction. Consequently it is to be expected that the indirect effects of surface compaction on plant growth will assume a greater relative importance in walking than in damage caused by running. Since the soil in amenity areas will also commonly be more vulnerable to compaction than sand-amended sports fields, it seems likely that a different kind of 'wear' simulation such as the walking cow's hoof of Scholefield & Hall (1986) will be more appropriate for studying the effects of walking.

Using the differential slip machine, Canaway (1979) has shown that abrasion rapidly reduced the ground cover of fine turf (*Festuca rubra*) from almost 100% to near 30% but that thereafter the decrease in ground cover was much less marked. Allcock (1973) obtained a similar form of relation between the standing biomass of grass on 5 cm thick turves and the number of impacts from a simple treading device. This suggests that, with suitable standardized treading, playing or wheel simulators, it is feasible to define the concept of a critical wear/treading threshold beyond which there is an unacceptable loss of ground cover. Such threshold values could then be related to the ground vegetation, soil type and soil physical conditions. In a later paper, Canaway (1983) studied the comparative tolerance of different grass species to football-type wear and found that *Poa annua* was much the most wear-tolerant species followed by *Lolium perenne*. However, the latter was considered preferable for football pitches because of the poor playing quality provided by *P. annua*.

When ground vegetation has been reduced by wear, compaction or other agencies, a competitive niche may become available to other species (see Ash, this volume).

EARTHWORMS

Earthworms play an important natural role in amelioration of soil physical properties and, in particular, in improving the surface infiltration rate. One to three hundred worm channels per square metre have been reported in pasture where, despite their negligible contribution to the overall porosity of the soil, they can play an important role in infiltration when the rainfall rate would otherwise have exceeded the maximum soil infiltration rate and led to runoff and possible erosion. Because the rate of water flow in a tube is proportional to the fourth power of its radius, a very few worm channels can have a large influence on infiltration. In common with the cracks in soils that shrink, worm channels, because of their comparatively large diameter, conduct water only when it is at virtually zero matric potential or under a slight positive pressure. This results in what has been called 'bypassing flow' in which the channels start to flow when the surface water is at virtually zero potential but this flow then bypasses the rest of the surface soil layer.

Worms play an additional role in soil mixing and in particular, in pulling detached fragments of organic matter down into their burrows where decomposition can take place at a faster rate. Where there is a low worm population (because of low soil pH, the use of a chemical worm killer, etc.) dead organic matter can accumulate on the soil surface

resulting in what is known (in grassland) as thatch. Whilst some accumulation of fibrous surface material is both natural and desirable (from the point of view of playing/walking quality) especially in fine sports turf, excessive amounts of thatch result in a surface that is too spongy and adversely affect playability. Thatch may be removed by scarification and controlled by improving aeration of the surface layers. The topic is reviewed by Shildrick (1985).

Baker (1981) has discussed the effects of earthworms in winter sports pitches. Despite the advantageous effects of worms, worm casts are a nuisance on the soil surface because they can interfere with the run of a ball and can result in a looser, wetter, muddier surface. He concludes that large areas of heavily worm-worked soil are undesirable on winter games pitches so that worm control measures may be necessary, supplemented by the mechanical worm-substitutes of spiking, to create artificial infiltration channels, and scarification.

STREET TREES

Literature on the physical properties of urban parkland (Craul, 1985; Short *et al.*, 1986) makes the point that urban subsoil may well present a poorly rootable medium with insufficient available water. In the case of trees that are surrounded by asphalt or paving stones, but whose root crown extends over a much larger area than that of the exposed soil at the base of the tree, the situation may be further exacerbated by one or more of a number of factors (Craul, 1985). First, the soil is likely to have been compacted except for a small volume surrounding the base of the tree, and road and building foundations may further restrict rooting; second, rainwater may be channelled on to the exposed soil in some sites causing waterlogging; third, there will be restricted oxygen access except through the uncovered soil; and finally, rainwater may run off the covered surface depriving the tree of sufficient water supply. Soil properties in the rooting zone of street trees have also been discussed by Brod *et al.* (1987).

Specialist topics: land reclamation

Several chapters in this book deal with different aspects of this diverse, multidisciplinary topic.

Physical problems that are likely to be encountered on reclaimed areas include: waterlogging, steepness of slope, stoniness, erosion risk and physical limitations to revegetation. Although some of these topics have already been considered, it is appropriate to add a few further comments.

In urban areas, the most common materials to be encountered on reclamation sites will be brickwastes and subsoils or various mixtures of the two. However, there will also be areas where colliery spoil, fly ash, sand and gravel, and a whole host of other byproducts of human activity are used as the basis for reclaimed soil. Thus the material on a reclaimed site should be considered as an artificial material to be characterized by its particle size distribution and those other properties listed in Tables 6.1 and 6.2 that are considered to be relevant.

According to Dutton & Bradshaw (1982) 'determinations of soil texture, particle density, bulk density and pore space are generally so variable even within a single site as to be meaningless' (in urban wastes). Although this is a forceful statement of the problem of soil heterogeneity, it should not be taken as a counsel of despair so much as an indication that site characterization may require considerable thought, effort and ingenuity. It is clear, for example, that a mixture of brickwaste and sandy soil may have radically different physical properties and problems to a mixture of brickwaste and clay. Thus the general information that may be obtained from a combined visual and auger survey (to detect hand texture, stoniness, waterlogging, etc. down the profile) may well provide a useful overview of the problems likely to be encountered in different parts of an heterogeneous site. In particular, it will usually be important to identify whether drainage problems are likely to exist, and if so, whether they are related to low hydraulic conductivity of the surface material, runoff, flow from broken drains coming from a neighbouring area, or to low conductivity of the soil at a greater depth. In each case a quite different solution is required.

Because the surface soil on most urban reclamation sites (excluding those on domestic waste or where a layer of topsoil has been added) will have little or no organic matter or structural development, soil management to allow the development of a surface layer capable of a high infiltration rate is an important consideration. Hackett (1977) indicates that the worst anticipated storm in a seven year period in the UK is usually taken as 37.5 mm of rainfall in 1 h. As discussed earlier, even material with <20% of particles <125 µm can have a hydraulic conductivity less than this rate. Thus, in most urban sites some runoff from areas under amenity use will be unavoidable during and after intense rainfall. More modest targets of soil performance that involve the maintenance of drainage conditions sufficient to avoid serious limitations on plant growth and to give trafficability/playability on most occasions are probably more appropriate.

A 'diagnostic' approach to urban amenity sites may be a worthwhile strategy. This would involve a detailed auger survey, probably with the additional use of dip-well or piezometers during a winter period to establish some idea of the site hydrology. This information will provide a guide both to the wetter and drier parts of the site as well as to the likely cause of the wetness. It should be possible to prepare a walk- or play-ability map of the site using similar criteria to those employed by the Soil Survey of England and Wales (e.g. Jarvis *et al.*, 1984). Effectively, these criteria amount to a rough estimate of how quickly a particular area of the soil will drain to a walkable/trafficable condition after prolonged heavy rainfall combined with climatic information on the dates at which the soil is likely to leave and return to field capacity in the spring and autumn. Although the Soil Survey have indicated how gley morphology may be used to characterize the water regime, this approach will not be possible on disturbed urban sites where the use of piezometers or dip-wells will be essential. An auger survey should also be able to indicate areas with low (i.e. droughty) and high profile available water.

Specialist topics: effluent disposal and renovation

Early treatment of sewage involved the direct application of untreated or partially treated sewage onto the land where the significant amounts of nitrogen and phosphorus it contained were utilized by growing plants. In such 'sewage farms', the sewage was applied by land filtration, broad irrigation or intermittent sand filtration (Escritt & Haworth, 1984). In each case, organic matter in the sewage was decomposed by aerobic microorganisms as it passed through or over the soil before the partially purified water reached water courses. This treatment fell into disrepute because of the possible transfer of toxic or otherwise undesirable sewage constituents to the farm vegetation and its consumers.

Modern sewage treatment can involve a variety of treatments including aerobic and anaerobic digestion, and mechanical dewatering (e.g. see Escritt & Haworth (1984) for more details). Sewage sludges are treated to render them more amenable for disposal and reduction in pathogens is incidental. Since pathogen content of sludges is very variable, origin and treatment of a sludge needs to be considered as well as a chemical analysis in deciding on its use on agricultural or reclaimed land. The topic of sludge disposal to land has been reviewed by a government working group (Hucker, 1981) and papers on the influence of sludge on soil physical and biological properties are given in Catroux *et al.* (1983). Hall (1988) has reviewed recent developments in methods of sludge application and their environmental impact.

Slurry (animal wastes) has to be stored during the winter in intensive farming systems and has also to be disposed of on the land in spring. This section is concerned only with the physical problems of the disposal of liquid effluents (sludge and slurry). Although emphasis has traditionally been placed on the disposal rather than the purification aspects of liquid movement, excellent disposal may be associated with poor purification due to short travel times so that both aspects need to be considered in defining an optimal application regime (Bouma *et al.*, 1983). An important limitation in applying water flow models is that concentrated effluent will not behave in exactly the same way as water because of its greater initial viscosity and because of possible pore clogging by organic solids. Sludge contains 2–7% dry solids and Arya (1985) quotes viscosities of 50–200 times that of water for such sludges.

An ideal soil for effluent disposal and renovation would allow avoidance of runoff or other forms of bypassing flow, adequate travel time, avoidance of compaction damage from wheel loads during application, as high a rate of application as desired, and no restrictions on timing of applications. Clearly these requirements are to some extent conflicting although a structureless sandy soil similar to that suggested as optimal for playing fields (but with a deep profile and no artifical drainage) probably represents the closest approach to satisfying all of these requirements. In practice, requirements one and two (and by implication three) must be taken as fundamental constraints that, taken together with the climate, weather and soil type determine the suitability, timing and appropriate application methods.

Where raw effluent is to be applied, odour is an additional constraint. Odours released from spreading of untreated sludge can persist for 1–3 weeks after spreading if sludge is left on the surface and are the principal cause of public complaints relating to sewage treatment and disposal (Hall, 1988). Injection techniques avoid this problem providing that the sludge is not injected faster than it can be accommodated beneath the surface. The EC directive on use of sewage sludge in agriculture (European Economic Community, 1986) requires all non-stabilized sludges to be immediately injected or worked into the soil following application. The enforcement of this directive from June 1989 therefore requires that all non-stabilized sludges applied to grassland are injected although there are, as yet, no comparable rules governing the larger quantities of slurry that are applied to land. Nevertheless, as injection results in reduced volatilization loss of ammonia in comparison to surface application and a significant increase in the proportion of nitrogen in the sludge or slurry being used by plants, there are also economic benefits to the use of injection on arable land (Hall *et al.*, 1987).

EFFECTS OF EFFLUENT ON SOIL PHYSICAL PROPERTIES

Sludge has a similar effect to other organic manures on soil physical properties. Many authors (e.g. Morel & Guckert, Hall & Coker, Guidi *et al.* (all in Catroux *et al.*, 1983)) have reported that sludge acts as a soil conditioner by increasing aggregate stability, hydraulic conductivity and organic matter content, and reducing bulk density. Pagliai *et al.* (1983) have also reported a reduction in the tendency for soil surface crusting after applications of sludge or pig slurry. Beneficial effects of sludge on physical properties have often been more pronounced on sandy and silty soils that would otherwise suffer from problems associated with low structural stability in the soil surface. Thus, where soil reclamation is involved, sludge may represent a valuable source of conditioner for an unstable soil.

Although the addition of sludge or other forms of organic matter is commonly reported to reduce bulk density and increase hydraulic conductivity, such results must be viewed in context. They usually refer to agricultural systems in which the organic solids are ploughed or otherwise incorporated into a loosened surface layer and where the bulk density and conductivity are representative of the more open packing state adopted by the more stabilized surface structure after the soil has had time to settle. Where effluent is regularly applied at high rates to the surface of uncultivated (e.g. grassland) sites there may be a decrease in hydraulic conductivity due to pore clogging (De Vries, 1972) or the build-up of a surface mat of organic matter with a low permeability to air and water that is detrimental to grass growth (Lea, 1979).

TIMING AND SOIL TYPE

Lea (1979) and Pollock (1979) have discussed the relation between the timing of slurry applications, weather and soil type. The same principles apply to the application of other organic effluents. The aim of controlling the timing of effluent applications is to avoid or minimize the possibility of some of the effluent flowing directly to drains or watercourses (by runoff or bypassing flow) and to maximize its travel time in the soil, thereby allowing sufficient time for decomposition and increasing the probability of deposition of the suspended solids. Therefore, as a general principle, applications should only be made when the soil is drier than field capacity and at such a rate that ponding and runoff cannot occur. Even under these circumstances it is important to note that: (i) if bypassing flow can occur then a proportion of any effluent applied may pass directly to the drains if a positive fluid pressure builds up (as observed by Pollock (1979)); (ii) where the SMD at the time of application is small, heavy rain shortly afterwards may flush some of the effluent into drains; and (iii) soil moisture deficit values are likely to be reliable when the soil has just left field capacity but may contain a considerable error later on in the growing season and therefore

do not provide a reliable guide to the date of return to field capacity in any given year.

It may be impossible to avoid applying slurry during the period in autumn, winter and spring when the soil is at, or wetter than, field capacity. Except after heavy rainfall, the soil still has some excess water storage capacity above the water table. This has been expressed in terms of the winter rain acceptance potential (WRAP). This is a semi-empirical rank ordering of soils effectively based on estimates of winter water table depth and the drainable porosity ($n - \theta[-5\,kPa]$), and a climatic factor (Farquharson *et al.*, 1978). Both Lea (1979) and Gardiner (1986) have suggested that WRAP may be used as a guide to the acceptibility of slurry applications during the period of zero SMD. However, considering the low rate of organic matter decomposition at winter temperatures, it seems desirable to check on the travel time of effluent applied to various soil types during winter and the applicability of the concept of WRAP as a guide to effluent disposal. Bouma *et al.* (1983) have compared a computer simulation model of effluent flow that allowed for bypassing flow through macropores with measurements in soil columns and *in situ* and obtained good agreement. In this way they were able to study travel time as a function of application rate. This kind of modelling approach should ultimately provide a means of achieving more precise criteria as to timing and rates of application in relation to soil type and soil water status.

METHODS OF APPLICATION

It is possible to use some conventional irrigation equipment (such as rainguns) for effluent application, to use a spreader attached to a tank, or to use an implement for soil injection. Because of the high axle loads when effluent tanks are full, they are capable of causing considerable compaction and damage to soil structure especially when used on soils close to field capacity. Use of rainguns that apply the effluent at a high rate may also cause breakdown of structure at the soil surface (Pollock, 1979) and hence risk runoff from soil with low structural stability or little vegetation cover.

Sludge injection is likely to prove an increasingly popular technique because it avoids odour, because of improved crop utilization of ammonia, and because pathogens in the sludge are buried beneath the surface. For this last reason it is thought acceptable that grazing be allowed three weeks after a successful injection (Hall & Davis, 1983). However injection may not be possible if the soil is too dry or stony and there is also some root damage involved. Hall & Davis (1983) have summarized a recent workshop on sludge injection. At that time there were seven commercially available injection systems in the UK, some capable of injection up to $170 \, m^3 \, ha^{-1}$. A design based on the winged subsoiler appears particularly promising because it is capable of creating a large cavity at a shallow depth to accommodate the sludge. Recent work in the UK (Hall *et al.*, 1986) has been incorporated into a code of practice on sludge injection (Water Research Centre/Silsoe College, 1986).

Arya (1985) and Arya *et al.* (1987) have designed injectors in which the pressure of the sludge is used to cause soil failure and reduce draught. Future work is needed on the relation between soil type and soil moisture status in relation to injection, with particular attention to the risks of incomplete burial and of bypassing flow to drains in structured soils.

REED BED/ROOT ZONE TREATMENT BEDS

In the past few years a new method of sewage treatment has appeared that involves continuously passing sewage along a gently sloping bed of soil or gravel planted with a bed of reeds (*Phragmites australis*). This system is appropriate for treatment of sewage from small towns and villages. Although 23 systems have been built in the UK, the system is still in the stages of active development and European research is now coordinated with the Water Research Centre acting as secretariat, who have recently summarized the current state of knowledge in the UK (Cooper *et al.*, 1988).

The principle of the system is that the rhizomes of the reeds open up the soil to provide a hydraulic pathway for flow and at the same time the reeds act as conducting pathways for atmospheric oxygen allowing aerobic bacterial decomposition to take place in their vicinity. In practice, systems are designed for hydraulic conductivities of $1-30 \times 10^{-4} \, m \, s^{-1}$ which can be achieved in gravel-filled systems but have not so far been reported for soil-filled systems where flow and decomposition often concentrate within a surface layer of organic debris. It is clear that, unless the rhizomes really can create an effective interconnected flow pathway, the necessary values of hydraulic conductivity can only be achieved in a well-structured soil (that may be difficult to maintain under waterlogged conditions), a gravel or coarse/medium sand with a very small percentage of fines, or some composite system. Despite this, some of the soil-based systems have given a satisfactory performance and a more detailed understanding of the dynamics of the bed hydrology may allow the design of more reliable and predictable systems.

Acknowledgements

I am grateful to the many people who have provided me with help and advice and in particular, Drs W.A. Adams, P. Campbell, J.E. Hall, J.A. Hobson, J.M. Hollis, R.J. Parkinson, D.L. Rimmer and D. Scholefield.

References

Adams, W.A. (1982). Planned use of turf-design and maintainance. *Mitteilingen Deutsche Bodenkundliche Gesellschaft* **33**, 215–223.

Adams, W.A. (1986). Practical aspects of sportsfield drainage. *Soil Use and Management* **2**, 51–54.

Adams, W.A., Stewart, V.I. & Thornton, D.J. (1971). The assessment of sands suitable for use in sportsfields. *Journal of the Sports Turf Research Institute* **47**, 77–85.

Adams, W.A., Tanavud, C. & Springsguth, C.T. (1985). Factors influencing the stability of sportsturf rootzones. In Lemaire, F. (ed.), *Proceedings of the 5th International Turfgrass Research Conference*, pp. 392–399. Avignon, France.

Allcock, P.J. (1973). Treading of chalk grassland. *Journal of the Sports Turf Research Institute* **49**, 21–28.

Arya, K. (1985). Soil failure by introducing sewage sludge under pressure. *Transactions of the ASAE* **28**, 397–400.

Arya, K., Tsunematsu, S. & Wu, L. (1987). A powered rotary subsoiler with pressurized sewage sludge injection. *Transactions of the ASAE* **30**, 1126–1130.

Avery, B.W. & Bascomb, C.L. (1974). *Soil Survey Laboratory Methods*. Technical monograph no. 6, Soil Survey, Silsoe, Bedford.

Avery, B.W. & Bullock, P. (1977). *Mineralogy of Clayey Soils in Relation to Soil Classification*. Technical monograph no. 10, Soil Survey, Silsoe, Bedford.

Baker, S.W. (1981). The effect of earthworm activity on the drainage characteristics of winter sports pitches.

Journal of the Sports Turf Research Institute **57**, 9–23.

Baker, S.W. (1983). Sand for soil amelioration: analysis of the affects of particle size, sorting and shape. *Journal of the Sports Turf Research Institute* **59**, 133–145.

Barnes, K.K., Carlton, W.M., Taylor, H.M., Throckmorton, R.I. & Vanden Berg, G.E. (Organising Cttee), (1971). *Compaction of Agricultural Soils*. ASAE, St Joseph, Minnesota.

Batey, T. & Childs, C.W. (1982). A qualitative field test for locating zones of anoxic soil. *Journal of Soil Science* **33**, 563–566.

Bell, M.J., Baker, S.W. & Canaway, P.M. (1985). Playing quality of sports surfaces: A review. *Journal of the Sports Turf Research Institute* **61**, 26–45.

Birse, E.L. (1971). *Assessment of Climatic Conditions in Scotland. 3. The Bioclimatic Sub-regions*. Soil Survey of Scotland, Macaulay Land Use Research Institute, Aberdeen.

Birse, E.L. & Dry, F.T. (1970). *Assessment of Climatic Conditions in Scotland. 1. Based on Accumulated Temperature and Potential Water Deficit*. Soil Survey of Scotland, Macaulay Land Use Research Institute, Aberdeen.

Black, C.E. (ed.), (1965). *Methods of Soil Analysis Part I, Physical and Mineralogical Properties Including Statistics of Measurement and Sampling*. ASAE monograph no. 9, Madison, Wisconsin.

Blackwell, P.S. (1988). The influence of soil geometry on soil compaction under wheels. In *Proceedings of the 11th International Soil Tillage Research Organisation Conference*, pp. 203–208. Edinburgh.

Bouma, J., Belmans, C., Dekker, L.W. & Jeurissen, W.J.M. (1983). Assessing the suitability of soils with macropores for subsurface liquid waste disposal. *Journal of Environmental Quality* **12**, 305–311.

Brod, H.G., Ellwart, T. & Hartge, K.H. (1987). Raumliche und zeitliche Anderung von Bodenparametern im Wurzelraum innerortlicher Alleebaume. *Mitteilungen Deutsche Bodenkundliche Gesellschaft* 55/II, 579–584.

British Standard 1377 (1975). *Methods of Test for Soils for Civil Engineering Purposes*. British Standards Institution. London.

British Standard 3969 (1965). *British Standard Recommendations for Turf for General Purposes*. British Standards Institution, London.

Canaway, P.M. (1976). A differential slip wear machine for the artificial simulation of turfgrass wear. *Journal of the Sports Turf Research Institute* **52**, 92–99.

Canaway, P.M. (1979). Studies on turfgrass abrasion. *Journal of the Sports Turf Research Institute* **55**, 107–120.

Canaway, P.M. (1982). Simulation of fine turf wear using the differential slip wear machine and quantification of wear treatments in terms of energy expenditure. *Journal of the Sports Turf Research Institute* **58**, 9–15.

Canaway, P.M. (1983). The effect of rootzone construction on the wear tolerance and playability of eight turfgrass species subjected to football-type wear. *Journal of the Sports Turf Research Institute* **59**, 107–123.

Castle, D.A., McCunnall, J. & Tring, I.M. (1984). *Field Drainage: Principles and Practices*. Batsford Academic. London.

Catroux, G., L'Hermite, P. & Suess, E. (1983). Proceedings of a Commission of the European Communities Seminar on *The Influence of Sewage Sludge Application on Physical and Biological Properties of Soils*. D. Reidel, Dordrecht, Holland.

Cavagna, G.A., Saibene, F.P. & Margaria, R. (1964). Mechanical work in running. *Journal of Applied Physiology* **19**, 249–256.

Cooper, P.F., Hobson, J.A. & Jones, S. (1988). Sewage treatment by reed bed systems: the present situation in the UK. *Paper presented to Institute of Water and Environmental Management, Birmingham*. Water Research Centre, Stevenage.

Craul, P.J. (1985). A description of urban soils and their desired characteristics. *Journal of Arboriculture* **11**, 330–339.

Darwin, C. (1882). *The Formation of Vegetable Mould through the Action of Worms*. John Murray, London.

Dexter, A.R. (1986a). Model experiments on the behaviour of roots at the interface between a tilled seed-bed and a compacted sub-soil. II. Entry of pea and wheat roots into sub-soil cracks. *Plant and Soil* **95**, 135–147.

Dexter, A.R. (1986b). Model experiments on the behaviour of roots at the interface between a tilled seedbed and a compacted sub-soil. III. Entry of pea and wheat roots into cylindrical biopores. *Plant and Soil* **95**, 149–161.

De Vries, J. (1972). Soil filtration of waste water effluents and the mechanism of pore clogging. *Journal of the Water Pollution Control Federation* **44**, 565–573.

Dutton, R.A. & Bradshaw, A.D. (1982). *Land Reclamation in Cities*. HMSO, London.

Escritt, L.B. & Haworth, W.D. (1984). *Sewerage and Sewage Treatment*. Wiley & Sons, Chichester.

European Economic Community (1986). Community directive on the protection of the environment and in particular of the soil when sewage sludge is used in agriculture. 86/278/EEC, *Official Journal of European Communities L181/6–12*.

Farquharson, F.A.K., Mackney, D., Newson, M.D. & Thomasson, A.J. (1978). Estimation of run-off potential of river catchments from soil surveys. *Soil Survey Special Survey no. 11*. Soil Survey, Silsoe, Bedford.

Farr, E. & Henderson, W.C. (1986). *Land Drainage*. Longmans, London.

Farrar, D.M. & Coleman, J.D. (1967). The correlation of surface area with other properties of nineteen British clay soils. *Journal of Soil Science* **18**, 118–124.

Gardiner, C.M.K. (1981). *The MORECS Discussion Meeting*. Report no. 78, Institute of Hydrology, Wallingford.

Gardiner, M.J. (1986). Use of soil and climatic data to predict hydraulic loading behaviour of Irish soils. *Soil Use and Management* **2**, 146–149.

Greacen, E.L., Barley, K.P. & Farrell, D.A. (1969). The mechanics of root growth in soils with particular reference to the implications for root distribution. In Whittington, W.J. (ed.), *Root Growth*, pp. 256–269. Butterworths, London.

Gupta, S.C. & Larson, W.E. (1979). A model for predicting packing density of soils using particle size distribution. *Soil Science Society of America Journal* **43**, 758–764.

Hackett, B. (1977). *Landscape Reclamation Practice*. IPC, Guildford.

Hall, D.G.M., Reeve, M.J., Thomasson, A.J. & Wright, V.F. (1977). *Water Retention, Porosity, and Density of Field Soils*. Technical monograph no. 9, Soil Survey, Silsoe, Bedford.

Hall, J.E. (1988). Methods of applying sewage sludge to land. A review of recent developments. In Dirkzwager, A.H. & L'Hermité, P. (eds), *Proceedings of the European Community Environmental Water Pollution Control Association Conference on Sewage Treatment and Use*, pp. 65–84. Elsevier Applied Science, Amsterdam.

Hall, J.E. & Davis, J.M. (1983). Sewage sludge injection into agricultural land. *Report of ICE workshop, London*. Water Research Centre, Medmenham, Bucks.

Hall, J.E., Godwin, R.J., Warner, N.L. & Davis, J.M. (1986). Soil injection of sewage sludge. *Report ER 1202-M*, 114 pp. WRc Environment, Medmenham, Bucks.

Hall, J.E., Godwin, R.J. & Warner, N.L. (1987). Soil injection of sewage sludge and animal slurries. In Welte, E. & Szabolcs, I. (eds), *Proceedings of the 4th International CIEC Symposium on Agricultural Waste Management and Environmental Protection*, pp. 285–292. Braunschweig.

Hodgson, J.M. (ed.) (1974). *Soil Survey Field Handbook: Describing and Sampling Soil Profiles*. Technical monograph no. 5, Soil Survey, Silsoe, Bedford.

Hodgson, J.M., Hollis, J.M., Jones, R.J.A. & Palmer, R.C. (1976). A comparison of field estimates and laboratory analyses of the silt and clay contents of some West Midland soils. *Journal of Soil Science* **27**, 411–419.

Hollis, J.M. (1987). *The Calculation of Crop-adjusted Soil Available Water Capacity for Wheat and Potatoes*. Research Report no. 87/1, Soil Survey, Silsoe, Bedford.

Hucker, T.W.G. (Chairman) (1981). *Report of the Sub-Committee on the Disposal of Sewage Sludge to Land*. National Water Council, London.

Jarvis, M.G., Allen, R.H., Fordham, S.J., Hazelden, J.,

Moffat, A.J. & Sturdy, R.G. (1984). *Soils and Their Use in South East England*. Soil Survey Bulletin no. 15, pp. 60–64, Soil Survey, Silsoe, Bedford.

Jones, R.J.A. & Thomasson, A.J. (1985). *An Agroclimatic Databank for England and Wales*. Technical monograph, no. 16, Soil Survey, Silsoe, Bedford.

Klute, A. (ed.) (1985). *Methods of Soil Analysis Part I: Physical and Mineralogical Methods*. 2nd Edition, American Society of Agronomy, Madison, Wisconsin.

Larson, W.E., Gupta, S.C. & Useche, R.A. (1980). Compression of agricultural soils from eight soil orders. *Soil Science Society of America Journal* **44**, 450–457.

Lea, J.W. (1979). Slurry acceptance. In *Soil Survey Applications*, pp. 83–100. Technical monograph no. 13, Soil Survey, Silsoe, Bedford.

Leeds-Harrison, P.B., Shipway, C.J.P., Jarvis, N.J. & Youngs, E.G. (1986). The influence of soil macroporosity on water retention, transmission and drainage in a clay soil. *Soil Use and Management* **2**, 47–50.

Luthin, J.N. (ed.) (1957). *Drainage of Agricultural Lands*. Monograph no. 7, American Society of Agronomy, Madison, Wisconsin.

Mackie-Dawson, L.A., Mullins, C.E., Fitzpatrick E.A. & Court, M.N. (1989). Seasonal changes in the structure of clay soils in relation to soil management and crop type. 1. Effects of crop rotation at Cruden Bay, NE Scotland. *Journal of Soil Science* **40**, 269–281.

Mandiringana, O.T. (1984). *The Relationship Between Soil Strength and Soil Water Tension in the Field*. MSc thesis, University of Aberdeen.

Marshall, T.J. & Holmes, J.W. (1988). *Soil Physics*. 2nd Edition, Cambridge University Press, Cambridge.

McKeague, J.A. & Topp, G.C. (1986). Pitfalls in interpretation of soil drainage from soil survey information. *Canadian Journal of Soil Science* **66**, 37–44.

McKeague, J.A., Wang, C. & Topp, G.C. (1982). Estimating saturated hydraulic conductivity from soil morphology. *Soil Science Society of America Journal* **46**, 1239–1244.

Ministry of Agriculture, Fisheries and Food (1967). *Potential Transpiration*. Technical Bulletin no. 16, MAFF, HMSO, London.

Ministry of Agriculture, Fisheries and Food (1983). *Technical Notes on Workmanship and Materials for Field Drainage Schemes*. BL4316. HMSO, London.

Mullins, C.E. & Fraser, A. (1980). Use of the drop-cone penetrometer on undisturbed and remoulded soils at a range of soil-water tensions. *Journal of Soil Science* **31**, 25–32.

Mullins, C.E. & Hutchinson, B.J. (1982). The variability introduced by various subsampling techniques. *Journal of Soil Science* **33**, 547–561.

Pagliai, M., Bisdom, B.A. & Ledin, S. (1983). Changes in surface structure (crusting) after application of sewage sludge and pig slurry to uncultivated agricultural soils in northern Italy. *Geoderma* **30**, 35–53.

Panayiotopoulos, K.P. & Mullins, C.E. (1985). Packing of sands. *Journal of Soil Science* **36**, 129–139.

Patto, P.M., Clement, C.R. & Forbes, T.J. (1978). *Permanent Grassland Studies no. 2. Grassland Poaching in England and Wales*. Grassland & Crops Research Institute, Maidenhead, Berks.

Pollock, K.A. (1979). The effect of slurry applications to soils. In *Soil Survey Applications*, pp. 101–109. Technical Monograph no. 13, Soil Survey, Silsoe, Bedford.

Reeve, M.J., Smith, P.D. & Thomasson, A.J. (1973). The effect of density on water retention properties of field soils. *Journal of Soil Science* **24**, 355–367.

Reeve, M.J., Hall, D.G.M. & Bullock, P. (1980). The effect of soil composition and environmental factors on the shrinkage of some clayey British soils. *Journal of Soil Science* **31**, 429–442.

Reid, I. & Parkinson, R.J. (1987). Winter water regimes of clay soils. *Journal of Soil Science* **38**, 473–481.

Robson, J.D. & Thomasson, A.J. (1977). *Soil Water Regimes*. Technical Monograph no. 11. Soil Survey, Silsoe, Bedford.

Scholefield, D. & Hall, D.M. (1985). A method to measure the susceptibility of pasture soils to poaching by cattle. *Soil Use and Management* **1**, 134–138.

Scholefield, D. & Hall, D.M. (1986). A recording penetrometer to measure the strength of soil in relation to the stresses exerted by a walking cow. *Journal of Soil Science* **37**, 165–176.

Schothorst, C.J. (1963). De draakracht van grasslandgronden. *Tijdschriffkoninklijke Nederlandsche Heidemaatschappij* **74**, 104–111.

Shildrick, J.P. (1985). Thatch: a review with special reference to UK golf courses. *Journal of the Sports Turf Research Institute* **61**, 8–25.

Short, J.R., Fanning, D.S., McIntosh, M.S., Foss, J.E. & Patterson, J.C. (1986). Soils of the Mall in Washington DC: I. Statistical summary of properties. *Soil Science Society of America Journal* **50**, 699–705.

Smalley, I.J. (1964). Flow-stick transition in powders. *Nature* **201**, 173–174.

Smith, D.L.O. (1987). Measurement, interpretation and modelling of soil compaction. *Soil Use and Management* **3**, 87–93.

Smith, K.A. & Mullins, C.E. (eds) (1990). *Soil Analysis: Physical Methods*. Marcel Dekker, New York.

Smith, L.P. & Trafford, B.D. (1976). *Climate and Drainage*. Technical Bulletin no. 34, MAFF, HMSO, London.

Soane, B.D. (ed.) (1982). *Compaction by Agricultural Vehicles: A Review*. Scottish College of Agricultural Engineering, Penicuik, Midlothian.

Spoor, G. & Ford, R.A. (1987). Mechanics of mole drainage channel deterioration. *Journal of Soil Science* **38**, 369–382.

Thompson, N., Barrie, I.A. & Ayles, M. (1981). *The Meteorological Office Rainfall and Evaporation Calculation System: MORECS*. Hydrological Memorandum no. 45. Meteorological Office, Bracknell.

van Shilfgaarde, J. (ed.) (1974). *Drainage for Agriculture*. Monograph no. 17, American Society of Agronomy, Madison, Wisconsin.

van Wijk, A.L.M. & Beauving, J. (1975). Relation between playability and some soil physical aspects of the toplayer of grass sportsfields. *Rasen-Turf-Gazon* **3**, 77–83.

van Wijk, A.L.M. & Beauving, J. (1978). Relation between soil strength, bulk density and soil water pressure head of sandy top-layers of grass sportsfields. *Zeitschrift fur Vegetationstechnik* **1**, 53–58.

Veihmeyer, F.J. & Hendrickson, A.H. (1948). Soil density and root penetration. *Soil Science* **65**, 487–493.

Vine, P.N., Lal, R. & Payne, D. (1981). The influence of sands and gravels on root growth of maize seedlings. *Soil Science* **131**, 124–129.

Wang, C., McKeague, J.A. & Topp, G.C. (1985). Comparison of estimated and measured horizontal K_{sat} values. *Canadian Journal of Soil Science* **65**, 707–715.

Ward, C.J. (1983). Sports Turf drainage: A review. *Journal of the Sports Turf Research Institute* **59**, 9–28.

Water Research Centre/Silsoe College (1986). *Soil Injection of Sewage Sludge. A Code of Practice.* 15pp. WRC, Medmenham, Bucks.

Webster, R. & Beckett, P.H.T. (1972). Matric suctions to which soils in south central England drain. *Journal of Agricultural Science, Cambridge* **78**, 379–387.

Whiteley, G.M., Utomo, W.H. & Dexter, A.R. (1981). A comparison of penetrometer pressures and the pressures exerted by roots. *Plant and Soil* **61**, 351–364.

Wild, A. (ed.) (1988). *Russell's Soil Conditions & Plant Growth*. Longman Scientific & Technical, Harlow, Essex.

Young, I.M., Costigan, P.A. & Mullins, C.E. (1988). Physical properties of two structurally sensitive soils and their effects on crop growth. In *Proceedings of the 11th International Soil & Tillage Research Organisation Conference* pp. 933–937. Edinburgh.

7 Nutrient provision and cycling in soils in urban areas

I. D. PULFORD

Introduction

A fundamental role of soil is to hold nutrients, and to release them in such a way that there is a steady supply to meet plant requirements. This view of soils was developed around the middle of the nineteenth century and led to their scientific study. By far the greatest attention in this work has been paid to agricultural soils and natural, but still essentially rural, soils (e.g. wetlands, moorland, forest soils). Soils in the urban environment have been somewhat neglected, although there has been considerable interest over the past 20 years or so in materials which result from particular industrial or mining processes (e.g. colliery spoil) (Bradshaw & Chadwick, 1980). One reason for this neglect of urban soils may be the wide and varied range of substrates for plant growth found in urban areas.

Plants require certain elements to ensure healthy growth. These essential elements are usually divided into two groups, depending on the relative amounts needed by most plants. Nutrients required in large amounts (macronutrients) are generally considered to be N, P, K, Ca, Mg and S, and the essential micronutrients are Fe, Mn, B, Cu, Zn, Cl and Mo.

The total amounts of nutrients in a soil often bear little relationship to those taken up by plants. What is important are the forms in which nutrients are held in soil, and the equilibria between various forms. Of particular importance is the balance between reactions which immobilize a nutrient (i.e. make it less available for plant uptake) and those which release it (i.e. make it more available for plant uptake). The soil may be viewed as a series of pools or reservoirs of nutrients, defined by the ways in which the nutrients are held. Commonly, eight such pools are considered: (i) soil solution; (ii) exchangeable ions, i.e. held by coulombic forces (physisorption); (iii) chelated by organic matter (both soluble and solid phase organic matter); (iv) incorporated into solid phase organic matter; (v) specifically adsorbed ions, i.e. held by covalent forces (chemisorption); (vi) precipitated compounds, i.e. controlled by solubility; (vii) biomass –

particularly in soil microorganisms and released on death and breakdown of the cell; (viii) soil minerals – generally very stable systems which release nutrients only slowly on weathering.

Plants growing in urban areas could be in a 'normal' soil (e.g. parks and gardens), or a 'degraded' soil (e.g. contaminated or derelict land) or a non-soil substrate (e.g. an industrial byproduct). Such materials can vary widely in nutrient content and in their ability to hold and cycle nutrients.

The information available on nutrient contents and nutrient cycling in soils of urban areas is limited and has tended to concentrate on contamination by products released due to industrial processes. Soils in urban areas have been neglected in comparison with soils used for agriculture because of their low economic importance in terms of plant production. As far as non-soil substrates are concerned, the greatest attention has usually been paid to problems of toxicity arising from specific industrial processes or land uses (e.g. gasworks sites, hydrocarbon contamination). However, there has been interest in the nutritional aspects of certain common, non-soil substrates. There is a considerable amount of information on coal-mine waste and strip-mine soils, as well as on certain other locally important wastes, such as oil shale and china clay wastes.

To understand the processes of nutrient provision and cycling in soils in urban areas, it is often necessary to extrapolate from information published for non-soil substrates.

Nitrogen

Most studies on degraded soils or industrial substrates conclude that the major nutritional limitation is a lack of nitrogen.

Nitrogen is only a very small component of the minerals constituting a soil or substrate. The main inputs of nitrogen are by biological tissues, which receive their nitrogen either from fertilizers or manures, or by microbial fixation of atmospheric nitrogen. Most of the nitrogen in a soil ($> 90\%$) is found as a component of the organic matter fraction (fresh or decaying biological tissue and humified organic matter). The small amount of inorganic nitrogen is present mainly as ammonium (NH_4^+) and nitrate (NO_3^-) ions. Soils have the capability to retain NH_4^+ by ion exchange, but usually have little retentive capacity for NO_3^-.

Most plants take up nitrogen as either NH_4^+ or NO_3^- and a continual supply is necessary to meet their requirements. Clearly, the small amounts normally found in a soil would be quickly exhausted by plant uptake unless there was some mechanism for their replenishment. Under suitable conditions, nitrogen is mineralized from the organic fraction by microbial action to the NH_4^+ form, which is then oxidized by other bacteria to NO_3^-. Other reactions which can occur under certain conditions are immobilization (microbial incorporation of N into the soil organic matter fraction), denitrification (loss of soil nitrogen as nitrous oxide or nitrogen gas) and volatilization (loss of soil nitrogen as ammonia gas). All of these reactions depend on there being suitable conditions of pH and aeration, a lack of toxic substances and, in many cases, a supply of organic matter as an energy source to sustain populations of the appropriate microorganisms. Because of the nature of the environment in degraded soils, contaminated soils or substrates, it is often the case that mineralization is inhibited or immobilization, denitrification and volatilization increased, resulting in a shortage of nitrogen available for plants.

QUANTITIES OF SOIL NITROGEN

Total nitrogen contents of soils range from $< 0.02\%$ in subsoils to $> 2.5\%$ in peats, with the surface layers of cultivated soils containing $0.06\%-0.5\%$ (Bremner & Mulvaney, 1982). In most soils nitrogen is found predominantly ($> 90\%$) as a component of the organic matter fraction (Stevenson, 1982a).

Nitrate is not held at the surfaces of soil components, except for minor amounts held by ion-exchange on positively charged sites on clays, oxides and organic matter, and so occurs predominantly dissolved in the soil solution. In many soils, especially well-managed agricultural soils, nitrate is the dominant anion in solution, being present at concentrations of approximately $10^{-2}-10^{-3}$M. In such soils conditions are suitable for the process of nitrification

to proceed, and so the levels of nitrate tend to be higher than those of ammonium. They also tend to be very variable, $2-20$ mg NO_3-N kg^{-1} soil being a typical range (Russell, 1973). Because nitrate ions are present only in solution they are readily available for uptake by plants, but are also rapidly leached.

Some of the NH_4-N is held by coulombic interactions at cation exchange sites on soil clays and organic matter. In this form it is considered to be available to plants, as it can easily be exchanged for another cation. This form of NH_4-N, however, constitutes only a small fraction of the ammonium ions in most soils. A far greater proportion is held as fixed ammonium, trapped between clay lattices and so unavailable to plants except by the slow process of weathering.

The meaning of measurements of the amounts of nitrogen in soils is difficult in relation to plant availability. Methods have been developed for total nitrogen (Bremner & Mulvaney, 1982), organic forms of nitrogen (Stevenson, 1982b) and inorganic forms of nitrogen (Keeney & Nelson, 1982). Supply of nitrogen to plants over a growing season depends on the release of nitrogen from organic matter, and the balance between this and losses of nitrogen.

Assessment of the amounts of nitrogen in urban soils is a meaningless exercise because, as mentioned above, the creation of an active nitrogen cycle is necessary to allow a continuing supply of N to plants. As most of the nitrogen in a soil is present in the organic matter fraction, it is clearly essential that this fraction is present to facilitate the cycling of nitrogen. In many urban soils, organic matter levels may be low due to either losses from topsoil by oxidation or erosion, or removal of topsoil leaving subsoil as the surface material. On undisturbed, vegetated sites, or in parks and gardens, amounts of organic matter may be large, especially under grass, due to a build-up with time. Many of the substrates which act as urban soils have very small organic matter, and hence nitrogen, contents (e.g. coal-mine waste, china clay waste).

Some of the total N measured in coal-mine wastes comes from the organic content of the black shale which is the predominant component of most wastes. It is unlikely, however, that this nitrogen plays any role in supply to plants, as it will be released only on breakdown of the shale. Reeder &

Berg (1977) measured total N levels of 730 mg N kg^{-1} in a fresh strip-mine spoil, 1083 mg N kg^{-1} in a vegetated spoil, 1112 mg N kg^{-1} in a Cretaceous shale and 1193 mg N kg^{-1} in a strip-mine soil from Colorado. Cornwell & Stone (1968) found total N in black shales from Pennsylvanian coal wastes in the range 3000–6900 mg kg^{-1}. Pulford (1976) obtained values of 1000–8000 mg total N kg^{-1} for coal-mine wastes of central Scotland. Values of 460–7000 mg total N kg^{-1} were measured by Aldag & Strzyszcz (1980) in ten phyllosilicates from the Polish coalfield.

In wastes with little or no organic component, the total N level is even smaller; 7–24 mg N kg^{-1} in acid molybdenum mill tailings (Berg *et al.*, 1975) and 350 mg N kg^{-1} in pulverized fuel ash (Townsend & Gillham, 1975). In china clay wastes, Dancer *et al.* (1977b) reported 36 kg N ha^{-1} in mica wastes and 18 kg N ha^{-1} in sand wastes, to a depth of 10 cm. Assuming a bulk density of 2 t m^{-3}, these figures are equivalent to 18 mg N kg^{-1} and 9 mg N kg^{-1} respectively.

Because little value can be placed on total nitrogen levels in relation to the availability of nitrogen to plants, attempts have been made by various workers to assess the amounts of nitrogen in specific fractions. Palmer *et al.* (1985) used the methods of Stevenson (1982b) to measure the various fractions of nitrogen in colliery spoil (Table 7.1).

As mentioned previously, fixed ammonium may be a quantitatively important component of the total N. Palmer & Chadwick (1985) measured 632 mg N kg^{-1} as non-exchangeable NH_4^+ in an unameliorated spoil at Thorne in South Yorkshire. This fraction represented 25.6% of the total N in the spoil. In a vegetated spoil at the same site, Palmer *et al.* (1985) reported 579 mg N kg^{-1} non-exchangeable N, this being 24.9% of the total nitrogen. Power *et al.* (1974) measured 150–300 mg N kg^{-1} as fixed ammonium in Paleocene shale. Aldag & Strzyszcz (1980) found 250–1155 mg N kg^{-1} of fixed nitrogen in phyllosilicate materials from the Upper Silesian Region of Poland. Young & Aldag (1982) quoted values of 3–10 mg fixed NH_4-N kg^{-1} for agricultural soils in the UK and USA, with 210–530 mg fixed NH_4-N kg^{-1} in Carbonaceous shales and 250–1540 mg fixed NH_4-N kg^{-1} in Carbonaceous phyllosilicates. Shah (1988) measured the ability of 90 coal-mine spoils from central Scotland to fix added ammonium by equilibrating spoil with 100 mg NH_4-N kg^{-1} at 2°C for 24 h, and then determining 0.5M K_2SO_4 extractable N. The difference between the amount of nitrogen added initially and that measured after 24 h represents a loss of N from solution by fixation. He reported a mean fixation of 6.0 mg N kg^{-1} with a range of 0–22.9 mg N kg^{-1}.

The forms of nitrogen which are immediately available to plants are nitrate and exchangeable and soluble ammonium. The amounts of these ions present in a soil are low and highly variable, depending on the balance between mineralization, or other gains, and immobilization, and other losses such as leaching. Aldag & Strzyszcz (1980) measured exchangeable NH_4^+ in ten phyllosilicates from coal spoils in Poland, and found amounts ranging from 10–119 mg N kg^{-1}. Shah (1988) measured 0.5M K_2SO_4 extractable NH_4^+ in 90 Scottish coal spoils and found values ranging from 0.1–25.9 mg N kg^{-1}, with a mean of 2.5 mg N kg^{-1}. Nitrate was detected in only 59 samples, and ranged from 0–9.0 mg N kg^{-1}, with a mean of 0.9 mg N kg^{-1}. The total extractable inorganic N ranged from 0.1–31.4 mg N kg^{-1}, with a mean of 3.4 mg N kg^{-1}. These are

Table 7.1 The major N fractions in freshly exposed, unameliorated colliery spoil from six sites in England and Wales and on a vegetated colliery spoil tip in Yorkshire (Palmer *et al.*, 1985), with typical soil values for comparison (Stevenson, 1982c)

Unameliorated colliery spoil	
Total N	2 212–4 316 mg N kg^{-1}
Incubated N	3.1–24.1 mg N kg^{-1} (0.1–0.6% of total N)
Hydrolysable N	448–1 490 mg N kg^{-1} (13.3–36.1% of total N)
Non-exchangeable N	338–1 685 mg N kg^{-1} (10.6–40.6% of total N)

	Vegetated colliery spoil	Typical soil values
	(% of total nitrogen)	
Non-hydrolysable N	67	20–35
NH_4-N in hydrolysate	28	20–35
Amino acid N	7	30–45
Amino sugar N	1	5–10
Hydrolysable unknown N	3	10–20

small values compared with 9.0–162.6 mg N kg^{-1} found in nine agricultural soils extracted under similar conditions (Khan, 1987). Shah (1988) found the largest concentrations of NH$_4$-N in the more acidic spoils, as did Williams & Cooper (1976) (Table 7.2).

INPUTS OF NITROGEN

Precipitation

Small amounts of nitrogen are added to soils in precipitation as NH$_4$-N, NO$_3$-N and organic forms of N. Although there may be localized variations in urban areas due to specific industrial emissions, studies on nitrogen in precipitation have shown generally similar amounts in different areas. Allen *et al.* (1968) measured total N in precipitation at five sites in northern Britain for periods up to three years. They found a range of N inputs to soil of 8.7–19.0 kg N ha^{-1} year^{-1} (mean 12.3). Dennington & Chadwick (1978) measured a range of total N of 8.46–9.58 kg N ha^{-1} year^{-1} added to three coal spoil tips in Yorkshire by precipitation. This input was made up of 4.1–4.6 kg NH$_4$-N ha^{-1} year^{-1}, 3.0–3.4 kg NO$_3$-N ha^{-1} year^{-1} and 0.9–1.5 kg organic N ha^{-1} year^{-1}. Tabatabai *et al.* (1981) reported that in the North Central Region of the USA precipitation added from 5–15 kg N ha^{-1} year^{-1}. At six sites in Iowa, the total N added in precipitation ranged from 9.8 to 14.3 kg ha^{-1} year^{-1} (mean 12.5), while at four sites in Minnesota the range was 1.6–11.6 kg N ha^{-1} year^{-1}, with a mean of 6.1.

Killham & Wainwright (1981) measured mean values of 200 mg kg^{-1} amino acid N and 4.7 mg kg^{-1} NO$_3$-N in five samples of atmospheric pollution deposits. They showed that this was sufficient nitrogen to support the growth of the fungus *Fusarium solani* in culture and in autoclaved and non-sterilized soils. This suggests that the nutrient content of atmospheric fallout could supply nutrients to soil microorganisms.

Fertilizers and manures

In unameliorated urban soils the input of nitrogen from fertilizers and manures will be minimal. It is only where reclamation of derelict or contaminated land is undertaken that additions of these materials will be made.

Fertilizer nitrogen may be added as an inorganic form, such as ammonium nitrate or ammonium sulphate. These are probably the commonest nitrogen fertilizers used in the UK, and may be added in these forms or in a compound with P and K fertilizers. Urea also forms a rapidly available form of nitrogen, providing that it is hydrolysed in soil. Slow-release forms of N fertilizer can be used to provide a continuing supply of N over the longer term. Compounds such as sulphur-coated urea and isobutylidene urea have been tested.

In a survey of different types of nitrogen fertilizer added to field trials on coal-mine waste, Palmer *et al.* (1986) found that ammonium sulphate was a much more effective form of nitrogen fertilizer than calcium nitrate. This was due to the retention of NH$_4$-N on cation exchange sites, while NO$_3$-N was easily and rapidly lost by leaching. In addition, they found that sulphur-coated urea produced the highest yields of herbage. Although the other slow-release nitrogen source tested, isobutylidene urea, gave poor yields at the first harvest, the yields improved at subsequent harvests. Palmer *et al.* (1986) concluded that slow release fertilizers were more suitable than the more readily available forms of nitrogen for the establishment of grass/clover swards on coal-mine waste.

Ascertaining the amount of nitrogen fertilizer which should be added to establish a grass/clover cover on urban soils and substrates involves a balance between adding a large amount to give the grass a good start (but risking a high degree of loss and inhibition of the legume component) and adding a smaller amount to provide an immediate supply to the grass, hoping that the legumes will initiate nitrogen fixation and that a self-sustaining system will develop and nitrogen contents gradually in-

Table 7.2 Extractable N in coal spoils of different pH (Williams & Cooper, 1976)

Locality	pH	2M KCl extractable (mg N kg^{-1})	
		NH$_4^+$	NO$_3^-$
Mitchell's Main	3–4	5–15	< 5
Roundwood	4–5	2.5–5	1–6
Upton	6–7	0–4	< 2.5

crease. Palmer (1987) has discussed the effect of amounts of nitrogen fertilizer on the establishment of clover on colliery spoil.

A positive response by vegetation to applied nitrogen has been observed in a variety of substrates; china clay wastes (Bradshaw *et al.*, 1975), urban demolition sites (Bloomfield *et al.*, 1982), surface-mined coal spoils (Ebelhar *et al.*, 1982) and colliery spoil (Bloomfield *et al.*, 1982; Pulford *et al.*, 1988). Dancer *et al.* (1979) showed that grass could be established on kaolin mining wastes (china clay) if nitrogen fertilizer was added. The yield of herbage in the first year was limited by the availability of N. If subsequent additions of N fertilizer were not made, the grass showed evidence of nutrient deficiency, there was a progressive reduction in herbage biomass and the nitrogen content of the herbage declined. A sward which had become moribund in this way showed rapid greening and regrowth on addition of nitrogen fertilizer. A linear response in herbage production to slow release fertilizers was found at rates up to $300 \, \text{kg} \, \text{N} \, \text{ha}^{-1}$. Shah (1988) reported a linear response in the yield of herbage dry matter on colliery spoil to application of nitrogen fertilizer up to $150 \, \text{kg} \, \text{N} \, \text{ha}^{-1}$ as ammonium nitrate. Visual observation of these plots indicated that white clover was inhibited at the higher rates of nitrogen addition.

Sewage sludge is the most readily available, and therefore most commonly used, organic amendment for urban soils and substrates. It can be obtained in a variety of forms, ranging from raw liquid sludge (approx. 4% dry matter) with typically 0.17% total N, to dewatered, digested sludge (approx. 50% dry matter) with typically 1.50% total N (ADAS, 1982). Sewage sludge has been widely used for land reclamation (Bradshaw & Chadwick, 1980; Sopper & Seaker, 1983). Many field trials using sewage sludge have been reported. Generally, sewage sludge applications have been found to be beneficial due to their nitrogen content, but any improvement tends not to be sustained beyond a few years without further additions. Problems of toxicity and run-off may occur due to over-application.

Olesen *et al.* (1984) used sewage sludge for the reclamation of abandoned brown coal pits in Denmark. They found that sludge applied at rates up to 120 tonnes dry matter per hectare enhanced grass growth. At the highest rate of addition, a grass cover of nearly 100% was maintained for six years. They did not recommend the use of sludge for tree establishment because the survival rate for pine seedlings decreased with increasing sludge application. Topper & Sabey (1986) studied the response of pasture vegetation to sewage sludge over two seasons on coal-mine spoil in Colorado. The spoil comprised mainly sandstone and shale parent material. Rates of addition up to $83 \, \text{t} \, \text{ha}^{-1}$ of dry sewage sludge were compared with a range of $0–160 \, \text{kg} \, \text{N} \, \text{ha}^{-1}$ as an inorganic N fertilizer. At rates of less than $83 \, \text{t} \, \text{ha}^{-1}$, the sewage sludge produced a larger vegetation yield than the inorganic fertilizer. The yield of grass with addition of $83 \, \text{t} \, \text{ha}^{-1}$ was similar to that at $160 \, \text{kg}$ $\text{N} \, \text{ha}^{-1}$ as inorganic fertilizer. The nitrogen content of the grass was enhanced at sewage sludge additions $> 14 \, \text{t} \, \text{ha}^{-1}$ (the smallest application rate used), compared to the plots treated with inorganic N. Pulford *et al.* (1988) reported results of an eight-year field trial on colliery spoil. They showed that sewage sludge at $20 \, \text{t} \, \text{ha}^{-1}$ increased the vegetation yield and nitrogen yield over the first four years, but that by the sixth year there was no difference between the treated and control plots.

Another frequently used organic amendment is poultry manure. This is particularly useful because of its high nitrogen content, much of which is readily available for plant uptake. ADAS (1979) gives typical values for the nitrogen contents of various types of poultry manure (Table 7.3).

Because of the large amounts of available nitrogen, care has to be taken in the application of poultry manure, as over-application can lead to scorching of the vegetation. If lime is also applied, very high pH values may result in the release of ammonia from the manure.

Palmer *et al.* (1986) used a composted chicken

Table 7.3 Nitrogen contents of poultry manure (ADAS, 1979)

	% Dry matter	%N	Approx. av. N $(\text{kg} \, \text{t}^{-1})$
Fresh droppings	25	1.4	9.0
Deep litter manure	70	1.7	10.0
Broiler manure	70	2.4	14.5
Dried poultry manure	90	4.2	25.0
In-house air-dried manure	85	4.1	24.5

manure applied to grass plots on colliery spoil at rates equivalent to 125 kg N ha^{-1} year^{-1} for two years (1975 and 1976), followed by 62.5 kg N ha^{-1} year^{-1} in 1978–1980. They found that this treatment was the poorest, compared to inorganic N fertilizers and slow-release N fertilizers, in terms of vegetation yield and nitrogen concentrations in the spoil. Pulford *et al.* (1988) added 4 t ha^{-1} chicken manure in field trials on colliery spoil and found that this gave the best results, compared to other organic amendments, in terms of vegetation yield and nitrogen yield in the herbage. After six years, however, the plots given chicken manure and the control plots were not significantly different.

Other organic materials which have been suggested as soil amendments include composted domestic refuse (Bradshaw & Chadwick, 1980) and shoddy (waste wool) (Williams & Cooper, 1976). Use of such materials as soil amendments is dependent primarily on their availability close to the site being reclaimed, as transport costs are usually the limiting factor.

Most of the studies on nitrogen supply to plants growing on degraded soils or substrates have been concerned with reclamation. In this context, the requirement very often is to establish a self-sustaining ecosystem which needs little or no after-care. This practice is determined by the economics of reclamation and contrasts with the more common consideration of soil in an agricultural system, where a continual supply of nitrogen is added to promote crop growth from year to year. Studies which have used inorganic N fertilizers or organic amendments, have tended to show that the applied nitrogen is sufficient initially, but that additional amounts must be applied to maintain nitrogen supply over the long term. Bradshaw (1983) suggested that a nitrogen capital of 1600 kg ha^{-1} would be required to provide a self-sustaining system. This is a higher figure than the 700 kg N ha^{-1} quoted by Dancer *et al.* (1977a). On the basis of this latter value, Bloomfield *et al.* (1982) stated that after-care maintenance would require 50 kg N ha^{-1} to be added for 15 years to provide the necessary nitrogen capital. This assumes no loss of nitrogen. If the higher figure of 1600 kg N ha^{-1} is adopted, the additions of N would have to continue for over 30 years. Clearly such treatments are impractical and uneconomic. Because of this, many workers have advocated the use of legumes to build up nitrogen resources in derelict soils and substrates.

Legumes

Many legume species have been used in land restoration, the commonest in Britain being: white clover (*Trifolium repens*), red clover (*Trifolium pratense*), alsike clover (*Trifolium hybridum*), birdsfoot trefoil (*Lotus corniculatus*), lucerne (*Medicago sativa*), tree lupin (*Lupinus arboreus*), white lupin (*Lupinus alba*) and gorse (*Ulex europaeus*)

To fix atmospheric nitrogen, the legume must be infected with the appropriate strain of the bacterium *Rhizobium*. Normally this would be ensured by use of inoculated seed. Experience with legumes on a variety of substrates has shown that, given suitable conditions for the legume and *Rhizobium*, significant amounts of nitrogen can be fixed, thus raising the nitrogen status of the soil.

Some estimates of the amounts of nitrogen fixed by legumes have been made using the acetylene reduction technique. Palmer & Iverson (1983) measured the rates of nitrogen fixation by white clover growing on colliery spoil. They reported values in the range 146–167 kg N fixed per hectare per year on plots which had received a large phosphate addition (98.4 kg P ha^{-1}), and 90–103 kg N fixed per hectare per year on plots which had received only 32.8 kg P ha^{-1}. No nitrogen fixation was measureable on plots not containing white clover. These values were considerably greater than those found by Skeffington & Bradshaw (1980), who measured a mean of 49 kg N fixed per hectare per year for white clover growing on china clay waste. They also measured rates of 72 kg N fixed per hectare per year by tree lupin and 26 kg N fixed per hectare per year by gorse, also on china clay waste. Otrosina *et al.* (1984) measured nitrogenase activity by the acetylene reduction method in kaolin spoil from Georgia and South Carolina. They found higher activities in spoil under herbaceous cover than on bare spoil or on newly established (seven year) pine trees. The highest activities were measured in spoil under well-established, undisturbed (30 year) pines, and in the naturally revegetated kaolin spoil. Fresquez *et al.* (1987), using the same technique, measured nitrogenase activity in an undisturbed soil and reclaimed coal-mine soils and spoils at a mine site in New

Mexico. The activity in the undisturbed soil was 204 n moles C_2H_4 produced per gram soil per day. On the reclaimed, non-topsoiled site the activity fell significantly at reclamation, but rose to 809 n moles C_2H_4 produced per gram soil per day after two years. The variation was such, however, that this value was not statistically different from that in the undisturbed soil. On the reclaimed, top-soiled site the nitrogenase activity was unaltered at reclamation, but rose to a maximum value of 1774 n moles C_2H_4 produced per gram soil per day after two years, falling thereafter to the previous level. These results suggest that nitrogenase activity can be built up from very low levels in a substrate such as coal waste by the establishment of a vegetation cover. Where a topsoil cover is used, the activity increases initially, presumably due to disturbance of the soil, but falls back to the previous level after about two years.

Where the amount of nitrogen fixed has not been measured by a technique such as acetylene reduction, assessments of nitrogen accumulation rates have been made by measuring the increase in N content of various components of the system with time. Dancer *et al.* (1977a) measured nitrogen accumulation rates in china clay wastes which had been naturally colonized. At the three sites studied, they determined mean rates of 63, 82 and 124 kg N ha^{-1} year^{-1} in the top 27 cm of spoil. The major plant colonizers were; tree lupin (*Lupinus arboreus*), broom (*Sarothamnus scoparius*) and gorse (*Ulex europaeus*). Average nitrogen accumulation rates in the plant biomass were 5, 20 and 21 kg N ha^{-1} year^{-1} respectively. Forage legumes planted on china clay waste showed even greater rates of nitrogen accumulation (Dancer *et al.*, 1977b). Starting from a very low nitrogen content of 36 kg N ha^{-1} for mica waste and 18 kg N ha^{-1} for sand waste, accumulation rates up to about 300 kg N ha^{-1} were measured over two years, (Table 7.4).

Marrs *et al.* (1980), also working on china clay waste in Cornwall, measured nutrient accumulation at 43 sites. They found mean nitrogen contents of 455 and 15 kg N ha^{-1} in reclaimed and raw mica dam walls, and 211 and 18 kg N ha^{-1} in reclaimed and raw sand tips respectively. The net accumulation rates for nitrogen were 67–263 kg N ha^{-1} year^{-1} (mean 130) for the mica dam walls, and −24 to 252 kg N ha^{-1} year^{-1} (mean 86) for the sand tips.

Table 7.4 Rates of nitrogen accumulation in china clay waste under different legumes (Dancer *et al.*, 1977b)

	Net annual N accumulation (kg N ha^{-1})	
	Mica waste	Sand waste
Trifolium pratense	102–281	119–151
Trifolium repens	137, 152	130, 135
Medicago sativa	140	115
Lotus corniculatus	210	117
Lupinus angustifolius	108	32
Vicia sativa	51	14

This study also considered the compartmentalization of nutrients between various components of the ecosystem (Roberts *et al.*, 1980) and showed that approximately 60% of the total nitrogen in the system was found in plant roots, 10% in plant shoots and 30% in the spoil. The net accumulation rates, in kg N ha^{-1} year^{-1}, on reclaimed, ungrazed systems were: mica dam walls; soil 39.2, roots 84.4 and shoots 14.4, and sand tips; soil 19.4, roots 61.6 and shoots 12.0. When calculating nitrogen accumulation rates in this way, some allowance must be made for leaching losses.

Palmer & Chadwick (1985) measured nitrogen accumulation over a period of seven years in grass swards growing on colliery spoil, where nitrogen was supplied by white clover or by ammonium sulphate. Annual accumulation rates were 54 kg N ha^{-1} on the white clover plots, 39 kg N ha^{-1} on the ammonium sulphate plots and 8 kg N ha^{-1} on the unamended plots. Over seven years, the fertilized plots received 500 kg N ha^{-1}, while the clover plots received no added nitrogen. Palmer & Chadwick (1985) reported that 67% of the nitrogen removed in the herbage on the clover plots was in the grass component. They suggested that almost all this nitrogen must have resulted from fixation by the clover, with subsequent transfer to the grass. Jefferies *et al.* (1981) had reached a similar conclusion regarding nitrogen transfer from legume to grass, and estimated that 76 kg N ha^{-1} was transferred from white clover to a companion grass within two years of sowing. Stroo & Jencks (1982) found a mean annual accumulation rate of 115 kg N ha^{-1} on sites on which black locust (*Robina pseudoacacia*), a leguminous tree, was growing. They attributed this high rate of accumulation of N to symbiotic fixation.

TRANSFORMATIONS AND CYCLING OF NITROGEN

In order that nitrogen levels can be built up, and nitrogen transformations and cycling take place, it is necessary for an organic matter fraction to be present in the soil. This acts as a reservoir for nitrogen, and as a source of energy for heterotrophic microorganisms. In undisturbed, vegetated soils or substrates it is likely that a suitable organic matter component will be present. Many substrates, such as mining wastes, may not support vegetation due to adverse conditions, and so an organic component must be developed. In a soil the process of nitrogen mineralization is a result of the breakdown of organic matter by microorganisms. Nitrogen is released along with carbon and other elements present in the organic matter. Ratios of N:C released are typically in the range 1:10–20 in temperate soils. Ammonium nitrogen may then be oxidized to nitrite and nitrate by the process of nitrification. In many urban soils and substrates, however, conditions may be such as to inhibit or prevent these transformations.

A major difficulty which is encountered when comparing the rates of nitrogen mineralization measured by various workers on a range of soils and substrates is the variety of methods used. The main variations are the use of fresh or air-dried sample, temperature of incubation and time of incubation. Because these factors differ from study to study, it is impossible to compare results directly.

Williams & Cooper (1976) used fresh colliery spoil which had been sieved to exclude material > 1 cm. Incubations were carried out at 15% moisture content at 27°C in loosely capped bottles, with evaporation losses being replenished with deionized water every two days. Extractable mineral nitrogen, using 2M KCl, was measured at 0 and 40 days. An unamended acid spoil (pH 4.5) mineralized 21 mg N kg^{-1} air-dried spoil over 40 days, while an unamended, neutral spoil (pH 6.9) mineralized only 8 mg N kg^{-1} air-dried spoil over the same period. When calcium carbonate was added to raise the pH to about 7.0, the two spoils mineralized 37 and 9 mg N kg^{-1} air-dried spoil respectively. Addition of organic amendments (shoddy and sewage sludge) resulted in more nitrogen being mineralized in the acid spoil. Where calcium carbonate was added, the mineral nitrogen was predominantly in the nitrate form, indicating that nitrification was proceeding. The effects of both the organic amendments and of calcium carbonate were much less marked in the spoil with an initial pH of 6.9.

Reeder & Berg (1977) measured carbon and nitrogen turnover in air-dried samples from a mine site in Colorado. The samples were sieved < 2 mm, rewetted to field capacity and incubated at 31°C for 168 days (Table 7.5).

Pulford *et al.* (1988) incubated fresh colliery spoil, which had received various treatments, at 50 kPa suction and 22°C. They measured carbon and nitrogen turnover as the slope of the linear part of the cumulative release plots (weeks 4–10 in a ten-week incubation) (Table 7.6).

Shah (1988) measured rates of nitrogen mineralization on 90 samples from spoil tips in central Scotland. He incubated fresh spoil which had been sieved < 4 mm at 50 kPa suction at 20°C. Mineral nitrogen was measured by 0.5M K$_2$SO$_4$ extraction, and carbon turnover as carbon dioxide released. He found nitrogen mineralization rates of 0–3.2 mg N per kilogram spoil per week (mean 0.8 mg N per kilogram spoil per week), and carbon turnover rates in the range 19.5–181.4 mg C per kilogram spoil per week (mean 62.6 mg C per kilogram spoil per week). On the same sites, he measured rates of nitrification of added ammonium (100 mg N kg^{-1} spoil). Only 46 of the 90 samples exhibited nitrification. The process was highly pH-dependent, being completely inhibited on sites of pH < 4.8. There were some sites of a higher pH which also showed no nitrification, suggesting a lack of colonization by the nitrifying microorganisms. In the samples in which added ammonium nitrogen was nitrified, the rate ranged from 2.1–71.4 mg N kg^{-1} week^{-1} (mean 25.2 mg N kg^{-1} week^{-1}).

Rates for carbon and nitrogen mineralization in agricultural soils have been widely reported. Typical values are 31.5–192.5 mg C per kilogram soil per week (Jenkinson & Powlson, 1976) and 1.7–4.2 mg N per kilogram per week (Tabatabai & Alkhafaji, 1980).

In general, carbon mineralization rates in sub-

Table 7.5 Carbon and nitrogen turnover in mine spoil (Reeder & Berg, 1977)

Materials	pH	Total N (mg N kg^{-1})	Nitrogen released (mg NO$_3^-$-N kg^{-1} spoil)	Carbon released (mg CO$_2$-C kg^{-1} spoil)	Nitrification of added NH$_4^+$ (mg NO$_3^-$-N kg^{-1} spoil)
Shale from active strip mine	7.4	1112	0	800	0
Vegetated spoil	7.9	1083	49	750	100
Fresh, non-vegetated spoil	8.0	730	5	500	36
Soil	7.2	1193	75	900	117

Table 7.6 Carbon and nitrogen turnover in coal spoil which received different treatments (Pulford *et al.*, 1988)

Treatment	mg CO$_2$-C released (kg^{-1} spoil week^{-1})	mg N released (kg^{-1} spoil week^{-1})
Chicken manure + lime	146.4	3.80
Sewage sludge + lime	127.5	2.72
Soil conditioner + lime	101.4	3.29
Peat + lime	143.6	2.48
No organic amd. + lime	102.0	3.46
No organic amd. no lime	16.5	0.68

strates tend to be similar to those measured in agricultural soils, whereas nitrogen mineralization rates are lower. This probably reflects the larger amounts of carbon relative to nitrogen which are available for microbial utilization in substrates and poor soils. This leads to a wide N:C ratio, which results in immobilization of nitrogen by microorganisms, leaving very little for plant uptake in the mineral nitrogen form. This situation is exacerbated by the addition of carbon-rich materials to soil, e.g. sawdust or hydrocarbons. It is well recognized for example that in soils contaminated with oil, nitrogen is rapidly immobilized by microorganisms (McGill, 1977). Losses of nitrogen from urban soils and substrates have not received much attention.

The lack of nitrogen supplied by mineralization may be due to some factor which results in the release of N being inhibited, the rapid loss of N due to microbial immobilization, fixation of NH$_4$-N by clay minerals or the lack of a sufficiently large pool of N. It is likely that on the types of materials being considered as urban soils or substrates, some or all of these factors may operate.

Bloomfield *et al.* (1982) studied nutrient supply in colliery shale and urban demolition sites (mainly brick waste) and concluded that the limiting factor was the lack of an adequate nitrogen capital in the soil. The same conclusion was reached in the study of china clay wastes in Cornwall (Marrs *et al.*, 1980; Roberts *et al.*, 1980; Skeffington & Bradshaw 1982). The main conclusion from these and other studies is that the most effective way of building up the nitrogen capital is by the use of legumes. In addition to raising the N content of the system, this is a successful way of increasing the level of mineralizable nitrogen. Jefferies *et al.* (1981) measured the nitrogen mineralized in reclaimed colliery spoil under patches of white clover, and in areas of no clover growth. After incubating the spoil at 30°C for ten days, they found that where clover was present a mean of 83.5 mg N kg^{-1} was released, compared to 4.8 mg N kg^{-1} at the non-clover sites.

Addition of an organic amendment can also affect the mineralization of N, both by stimulation of microbial activity and by increasing the nitrogen content. Voos & Sabey (1987) added sewage sludge at rates of 0, 40, 80 and 120 t ha^{-1} to stockpiled topsoil, undisturbed topsoil, mine spoil, a mixture of stockpiled soil and mine spoil, a mixture of undisturbed topsoil and mine spoil, and sand. They measured the nitrogen mineralized over 16 weeks at 25°C in samples at field capacity. In all cases the smallest amount of mineralized nitrogen was in the sample which had received no sewage sludge, and the largest rate in the sample treated at $120\,t\,ha^{-1}$. The total amounts of nitrogen mineralized over 16 weeks (in $mg\,N\,kg^{-1}$) are shown in Table 7.7. All of the soils and substrates had a pH > 6, and yet the sand showed no evidence of nitrification, while this process occurred in the mine spoil to only a limited degree. In the soils and soil/spoil mixtures, nitrification did occur, suggesting the presence of suitable organisms in these materials only.

For nitrogen transformations and cycling to occur in soils and substrates, it is necessary for a varied microbial population to become established. One way in which this can be measured is by monitoring respiration rates, either as oxygen consumed or as carbon dioxide released. A second way is to measure the activities of various enzymes. The activities of carbohydrases such as amylase, cellulase and invertase provide a guide to the breakdown of plant tissue in the soil or spoil, and hence an assessment of the development of an organic component. In relation to nitrogen transformations specifically, the activities of urease and amidase have been measured.

Stroo & Jencks (1982) measured the activities of amylase and urease in barren minesoils, minesoils with black locust cover, minesoils with a grass–legume cover and native soils, all in West Virginia.

Table 7.7 Effect of sewage sludge addition on nitrogen mineralization in soil and coal spoil (Voos & Sabey, 1987)

Material	Nitrogen mineralized over 16 weeks $(mg\,N\,kg^{-1})$
Stockpiled topsoil	392–938
Undisturbed topsoil	636–1023
Mine spoil	274–617
Stockpiled topsoil and mine spoil	319–680
Undisturbed topsoil and mine spoil	359–730
Sand	40–273

Amylase activity increased with age in the black locust soils, ranging from approximately $2\,\mu$ moles of reducing sugar per 100 g soil per hour in sites of about ten years in age, to $3.5\,\mu$ moles of reducing sugar per 100 g soil per hour in 20-year old soils. In the soils with a grass–legume cover, however, amylase activity (expressed as micromoles of reducing sugar per 100 g soil per hour) fell with age from 6.5 in 3-year old sites to 3.6 at 17 years. These values compared with rates of 0.6–1.0 in the barren soils and about 5 in the adjacent native soils. Amylase activity was highly correlated with microbial respiration rate and mineralizable N. Urease activity, however, was not correlated with these parameters, and showed no obvious pattern but was similar in all soils. Values were in the range 6.8–23.6 µg urea hydrolysed per gram soil per hour (3.2–11.0 mg NH_4-N released per kilogram per hour). Stroo & Jencks (1982) suggested that on unamended sites microbial activity increased with the establishment of vegetation, and hence in the accumulation of organic matter and nitrogen. On amended sites, it was possible to have high respiration rates with low to moderate levels of organic matter and nitrogen. Thus it seems that these minesoils could decompose added residues without showing a permanent increase in microbial activity on intensively managed sites. This reflects a common finding that vegetation on reclaimed sites, while good initially, may die off once the added nitrogen has been used up or lost, due to ineffective cycling or poor retention. This highlights the problem of nitrogen supply, which can be dealt with either by the use of legumes, or by suitable land management such as grazing or cropping.

Fresquez *et al.* (1987) measured a variety of enzyme activities in undisturbed, topsoiled and non-topsoiled sites in New Mexico, where soils are generally alkaline and often saline-sodic in nature. The soils in this area had pH values in the range 6.7–7.8, compared to those of 3.5–7.3 in the study of Stroo & Jencks (1982). Fresquez *et al.* (1987) measured the activities of amylase, cellulase and invertase. The activities of these enzymes (in micrograms glucose released per gram of soil per day) in the undisturbed soil were 131, 52 and 1225 respectively. On both topsoiled and non-topsoiled sites the activities fell at revegetation, but increased to levels not significantly different to the undisturbed soil within two to four years. The effect was less on the topsoiled sites compared to the non-topsoiled sites. Urease activity in the undisturbed soil was 834 µg NH_4-N released per gram soil per day. In the non-topsoiled site, this fell to 483 at revegetation, but rose to 802 within two to four years. Statistically, however, there was no difference between these values. In the topsoiled site, the activity fell to 680 µg NH_4-N released per gram of soil per day at revegetation, peaked at 2710 after two years, and fell to 983 by the fourth year.

Shah (1988) measured the activities of urease and amidase in 30 samples from experimental plots which had received various amendment treatments at a reclaimed coal mine waste site in central Scotland. The methods used to measure enzyme activity were the buffer methods of Tabatabai and Bremner (1972) for urease, and Frankenberger and Tabatabai (1980) for amidase. Enzyme activities in coal spoil and agricultural soils are given in Table 7.8.

Phosphorus

Unlike nitrogen, phosphorus may be a constituent of the minerals found in many soils and substrates. The amount of phosphorus supplied from this source tends to be small, and additions of P are usually essential for a continuing supply to plants.

The reactions of phosphorus in soils tend to result in it being largely unavailable to plants. This may be due to the insolubility and slow weathering of the phosphorus-containing minerals, or the rapid adsorption reactions which orthophosphate ions, released from minerals or added to soil, undergo. Some phosphorus is held as a constituent of the soil organic matter and phosphate is released by the process of mineralization. The amount of P held in this fraction is variable and depends on the organic matter content of the soil. Phosphate released by mineralization is usually in the ratio of N:P of 10:1 to 10:2 (Wild, 1988).

The general conclusion is that phosphorus in urban soils and substrates is unavailable to plants, and so plants suffer from a deficiency. There does appear, however, to be some discrepancy between an assessment of soil P status made using one of the standard extractants, and the response of plants to added phosphorus.

QUANTITIES OF SOIL PHOSPHORUS

Phosphorus is found in soils in amounts varying between 500 and 2500 kg ha^{-1}, with 15–70% in an inorganic form (White, 1979). The content is variable, the inorganic component depending on the nature of the soil minerals and on inputs of P fertilizers, the organic component on the amount of organic matter present and its rate of accumulation. Plant available phosphorus is usually assessed by extraction, using one of a wide range of extractants which have been developed to measure phosphorus availability in soils of varying properties (Olsen & Sommers, 1982).

The major form of phosphorus in soil is the orthophosphate anion; $H_2PO_4^-$ being the dominant form up to about pH 7, above which HPO_4^{2-} predominates. $H_3PO_4^0$ and PO_4^{3-} species are unimportant in soils, except under conditions of extreme acidity or alkalinity. The concentration of phosphate in soil solution is low, typically being of the order of 10^{-6}M. The main reason for this low amount of P in solution is the rapid adsorption of orthophosphate ions on to surfaces of hydrous oxides.

Quantities of phosphorus which have been measured in urban soils and substrates are highly vari-

Table 7.8 Urease and amidase activities in coal spoil and agricultural soils

Material	Enzyme activity (mg NH_4-N kg^{-1} h^{-1})	
	Urease	Amidase
Coal spoil (Shah, 1988)	3.9–33.8	3.0–41.4
Soil (Tabatabai, 1977)	9.0–131.3	–
(Frankenberger & Tabatabai, 1980)	–	97.5–167.5
(Frankenberger & Tabatabai, 1982)	16.5–110.0	52.5–224.5
(Zantua & Bremner, 1975)	3.0–109.1	–

able. Furthermore, a wide range of extractants has been used by different workers in an attempt to quantify specific fractions of the soil P. Amounts of P in coal-mine waste tend to be low. Palmer (1978) used a number of standard P extractants on spoil from the Yorkshire coalfield, and found that the amounts of P removed were too close to zero to be of value. Fitter *et al.* (1974) reported a mean value of $5.5 \, mg \, P \, kg^{-1}$ for bicarbonate extractable P in Lancashire coal spoil. On five unreclaimed sites in South Lancashire, Fitter & Bradshaw (1974) measured values of various pools of P. Their results showed ranges of $1.78–4.41 \, kg \, P \, ha^{-1}$ $NaHCO_3$ extractable P, $0.76–4.14 \, kg \, P \, ha^{-1}$ anion exchange resin extractable P and $13.4–71.1 \, kg \, P \, ha^{-1}$ Al + Fe bound P, (assuming 1 ha soil to 15 cm weighs 2240 t). Using spoil from the Scottish coalfield, Pulford (1976) found no 0.5M acetic acid extractable P in acid spoil. On the other hand, Lister (1987), using the Bray method (0.1M HCl + 0.05M NH_4F) also on Scottish spoil, measured extractable P from 0 to $19.3 \, mg \, P \, kg^{-1}$, with a mean of $6.2 \, mg \, P \, kg^{-1}$.

Gemmell & Goodman (1980) measured a mean of $70 \, mg \, P \, kg^{-1}$ 0.5M acetic acid extractable P in unamended zinc smelter waste in the Swansea Valley. Waste treated with domestic refuse contained $156 \, mg \, P \, kg^{-1}$, with sewage sludge $140 \, mg \, P \, kg^{-1}$ and with pulverized fuel ash $135 \, mg \, P \, kg^{-1}$. Pulverized fuel ash typically contains relatively large amounts of phosphorus. Townsend & Gillham (1975) quote approximately $630 \, mg \, P \, kg^{-1}$ as total P, and an average of $144 \, mg \, P \, kg^{-1}$ as 'available' P. Bradshaw & Chadwick (1980) gave a figure of $94 \, mg \, P \, kg^{-1}$ as a typical available content of pulverized fuel ash. Berg *et al.* (1975) measured extractable P in acid molybdenum tailings by both the Bray method (0.03M NH_4F/0.025M HCl) and $NaHCO_3$ extraction. The amounts extracted were in the ranges $2–11 \, mg \, P \, kg^{-1}$ and $1–10 \, mg \, P \, kg^{-1}$ respectively. These values indicate a deficiency of P compared to the levels commonly found in soils. China clay waste is generally deficient in nutrients. Marrs *et al.* (1980) measured total P in china clay waste from Cornwall by wet ashing with 4:1 HNO_3:$HClO_4$. They found $60 \, kg \, P \, ha^{-1}$ in raw mica dam walls and $79 \, kg \, P \, ha^{-1}$ in sand tips. These are very low values, and much of this P will be unavailable.

As with nitrogen, the usefulness of extractable P in substrates is doubtful. The many extractants used for soils were a result of the different ways in which

phosphate is held in soils of differing properties. A good extractant for an acid soil, for example, may not be suitable for an alkaline soil. Berg (1978) addressed these problems with regard to disturbed soils. Dancer (1984) examined the use of a number of standard soil tests for plant available phosphorus on newly reclaimed alkaline (pH 7.25) mine spoil, particularly in relation to phosphorus additions for maximum maize yield. He found that the mine spoils needed twice the amount of extractable P compared to topsoil to support maximum maize production. This suggests that assessments of sufficiency of phosphorus in soils, as determined by extraction, may not be directly applicable to mine spoils or other substrates. The concept of nutrient availability in substrates may differ from that in true soils, because adsorption and desorption mechanisms for phosphate may be substantially different. Fitter (1974) has suggested that the phosphorus-buffering capacity of spoils should also be considered when assessing phosphorus requirements. Further evidence for this view comes from studies on phosphate behaviour in spoils and plant response experiments (see later).

INPUTS OF PHOSPHORUS

Precipitation

Amounts of phosphorus in precipitation are small, unlike nitrogen and sulphur, as no P-containing gaseous compounds are released into the atmosphere. The major source of atmospheric P is particulate matter, mostly dispersed dust and soil. Allen *et al.* (1968) found only $0.2–1.0 \, kg \, P \, ha^{-1} \, year^{-1}$ (mean 0.54) deposited in precipitation at five sites in northern Britain. Similar annual deposition rates were measured by Dennington & Chadwick (1978) at three sites in Yorkshire, $0.22–0.35 \, kg \, P \, ha^{-1} \, year^{-1}$, and by Tabatabai *et al.* (1981) in the north central region of the United States, $< 0.1–1.0 \, kg \, P \, ha^{-1} \, year^{-1}$.

Fertilizers and manures

Some phosphorus will be released into urban soils and substrates by weathering of P-containing minerals and mineralization of organic matter. In established, vegetated soils and substrates it is likely that phosphorus supply is in equilibrium with plant uptake, but in many cases a response to added P is probable. When reclamation is undertaken, phosphorus is invariably added as a fertilizer. This is often triple superphosphate, which is a concentrated and cost-effective form. Manures may also add significant amounts of P. Sewage sludges contain total phosphorus contents of between 0.15% P_2O_5 in liquid raw sludge and 1.10% P_2O_5 in dewatered, digested sludge. Available amounts are 0.8 kg P_2O_5 m^{-3} and 5.5 kg $P_2O_5 t^{-1}$ respectively (ADAS, 1982). Poultry manure contains even higher amounts (ADAS, 1979) (Table 7.9).

TRANSFORMATIONS AND CYCLING OF PHOSPHORUS

The reaction which dominates the behaviour of phosphate added to soils or substrates is usually surface adsorption by hydrous oxides. In calcareous materials, adsorption by calcium carbonate surfaces and precipitation of calcium phosphates may be important processes. These reactions are rapid and extremely effective in removing phosphate ions from solution.

Pulford & Duncan (1975) and Doubleday & Jones (1977) both discuss phosphate adsorption by colliery spoil, and in particular the influence of ignition and acidity. Both sets of authors concluded that amorphous iron oxide surfaces were the most important for phosphate adsorption. Some colliery spoil tips ignite spontaneously, resulting in a red, as opposed to black, shale. The process of burning decreases the amount of amorphous iron oxide (as measured by acid oxalate extraction) and phosphate adsorption

(Pulford & Duncan, 1975). This occurs when the temperature exceeds 450°C, and may be due to crystallization of the amorphous iron oxides and fusion of the clay-sized materials. Both of these processes lead to a decrease in surface area. In pyritic coal spoil, iron oxide is a product of the process of pyrite oxidation, and when formed is amorphous, has a high surface area and so is highly reactive. Again, both sets of authors showed that there was an increase in P adsorption by coal spoil as the pH declined. This may be partly attributed to the formation of iron oxide, but also to the increase in positive charge at pH-dependent sites, making it easier for phosphate to approach the surface. There may also be precipitation of iron and aluminium phosphates under acid conditions.

Little work has been done on phosphate adsorption by other substrates, but a vast amount of information is available on the process in soil (e.g. White, 1980).

Various ways have been used to measure the phosphate-adsorption capability of a soil or substrates. A quantity/intensity (Q/I) plot, which relates the amount of P removed from solution in a given time with the concentration of P remaining in solution, is the most appropriate. From this, an estimate can be made of the soil's ability to release adsorbed P as the solution P concentration decreases (phosphate-buffering capacity). To obtain a value for the maximum amount of phosphate which a soil or substrate can adsorb, such Q/I data can be fitted to the Langmuir equation (White, 1980). Often, however, single point measurements of P adsorption have been used in preference to a multipoint isotherm. Fitter (1974) used the method of Ozanne & Shaw (1967), which determines the phosphorus requirement of a soil as the amount of P held by the soil at an equilibrium solution concentration of 0.3 mg $P l^{-1}$ (approximately 10 µM). Pulford & Duncan (1975) equilibrated 150 µg $P g^{-1}$ spoil in a system buffered at pH 4.6.

Mineralization of organic forms of phosphorus is important, occurring along with C and N release in the ratio of approximately 100 C:10 N:1 P (Wild, 1988). The contribution of this pool of P in degraded soils and substrates has been largely ignored, and even in agricultural systems its importance is not always appreciated. A study of P mineralization in Scottish coal spoils failed to show net release of phosphate, although this may be due to rapid

Table 7.9 Phosphorus contents of poultry manure (ADAS 1979)

	% Dry matter	% P_2O_5	Approx. available P (kg $P_2O_5 t^{-1}$)
Fresh droppings	25	1.1	5.5
Deep litter manure	70	1.8	9.0
Broiler manure	70	2.2	11.0
Dried poultry manure	90	4.3	21.0
In-house air-dried manure	85	3.1	15.5

adsorption of any mineralized P by hydrous oxides (T. A. B. Walker, unpublished results). Some estimates have been made of phosphatase enzyme activity in mine soils and spoils. Stroo & Jencks (1982) measured acid phosphatase activities (buffered at pH 6.5) in minesoils from West Virginia, and found values ranging from 0.6 to 4.8 μ moles (83–667 μg) *p*-nitrophenol released per gram of soil per hour. These values compared with 5.1 and 8.6 μ moles (709 and 1195 μg) *p*-nitrophenol released per gram of soil per hour for adjacent native soils. The phosphatase activities in the minesoils increased with time after reclamation and were related to organic matter and nitrogen accumulation. The activities in the minesoils were always lower than in native soils, even 20 years after reclamation when vegetation was well established, suggesting a slower rate of P mineralization in the minesoils.

Marrs *et al.* (1980) assessed net P accumulation rates on reclaimed china clay wastes. They reported ranges of 12–288 kg P accumulated per hectare per year (mean 54) in mica dam walls, and −50 to 906 kg P accumulated per hectare per year (mean 113) in sand tips. They estimated an overall annual loss of approximately 4 kg P ha^{-1}, taking the difference between added P and the measured accumulation rate. This gives a capture efficiency for P of 86%, which is higher than natural ecosystems, but lower than intensively managed agricultural systems. Marrs & Bradshaw (1980) reported leaching losses of P from china clay wastes in lysimeters of 4 kg P ha^{-1} year^{-1} in raw waste with developing vegetation, and 19 kg P ha^{-1} year^{-1} in vegetated waste. Much of the phosphate added to china clay wastes remains in the soil component. Roberts *et al.* (1980) found that, of the total P content of 181 kg ha^{-1} in mica dam walls, 82% was in the soil component, 14% in plant roots and 4% in shoots. Similar figures were measured for the sand waste.

Phosphorus availability to plants may be increased by infection with mycorrhizal fungi (Daft & Hacskaylo, 1976) or by stimulation of microbial activity (Fresquez & Lindemann, 1983).

RESPONSES TO ADDED P

Plant response to added phosphate has been measured on coal-mine waste and china clay waste. It is generally accepted that these substrates are deficient in P, due to the low amounts present and, in the case of some coal spoils, high P adsorption capability. A response to phosphate has often, but not always, been found in these substrates. One reason for this discrepancy may be due to the way in which the response experiment was carried out; pots in a glasshouse may give different results to field plots.

Fitter & Bradshaw (1974) tested the effect of N, P, K and lime additions in factorial experiments both in the glasshouse and in the field. They found highly significant responses by *Lolium perenne* to added P in all five spoils tested in the glasshouse experiment. In the field trial, one of the four spoils tested showed no response by *Lolium perenne* to added P, while the significance of the response in the other three spoils was less than in the glasshouse experiment. For *Agrostis tenuis*, only two of the four spoils showed a significant response to P in the field. These observations may have been due to greater variability in the field, or to some other factor influencing uptake of P. Doubleday & Jones (1977) and Lister (1987) showed significant responses to added phosphate fertilizer by *L. perenne* growing in pots. In a growth chamber study, Pulford & Duncan (1978) measured a small decrease in yield of *L. perenne* growing in an acid coal-mine waste and an old, vegetated coal-mine waste, but not in unvegetated waste, when a nutrient solution with no P was added compared to use of a complete nutrient solution.

Ebelhar *et al.* (1982) studied the effects of amendments added to surface-mined coal spoils, and found no significant effect of P on yield of common bermudagrass. Phosphate addition did help to increase winter hardiness and survivability. Pulford *et al.* (1988) found no effect on yield of herbage in a grass – clover sward of additions of 30 or 60 kg P ha^{-1} to field plots.

Discrepancies in plant response between pot and plot studies may arise due to differences in soil volume which can be exploited by roots, differences in intensity of leaching of the soil or changes in microbial activity. A further aspect to be considered where legumes are present is the stimulatory effect of P on their growth, and hence on nitrogen fixation. A response by plants in this case may be due primarily to the additional N fixed by the legumes.

The supply of P, and the response of plants to P, in urban soils and substrates has been studied to a lesser degree than for nitrogen, and is clearly less well understood. Phosphorus supply may, however, be second only to N supply in importance to the nutrition of plants growing in such soils.

Other macronutrients

Of the elements other than N and P defined as macronutrients, K, Ca and Mg exist in soils as cations, while S is present in aerated soils as the SO_4^{2-} anion, although under extremely anaerobic conditions the S^{2-} form may occur.

Potassium, calcium and magnesium are all commonly found as components of the minerals which constitute the parent materials of many soils and substrates. As weathering proceeds these ions are released, and can be held on cation exchange sites in forms available to plants. In general, the supply of these elements to plants in urban soils or substrates does not present any problems.

Small amounts of K, Ca and Mg are added by precipitation. Allen *et al.* (1968) found mean values of 4.0 kg K ha^{-1}, 11.4 kg Ca ha^{-1} and 4.2 kg Mg ha^{-1} deposited per year at five sites in northern Britain. Dennington & Chadwick (1978) measured similar annual deposits on three spoil sites in Yorkshire; 2.0–5.8 kg K ha^{-1}, 11.1–23.0 kg Ca ha^{-1} and 2.5–5.4 kg Mg ha^{-1}.

The amounts of Ca, Mg and K extracted from various substrates are shown in Table 7.10. These figures show the variability of such materials.

Doubleday (1972) and Pulford & Duncan (1978) found a significant decrease in available K in colliery spoil as pH decreased. This decrease is due to displacement and leaching of exchangeable K by H$^+$ ions. Pulford & Duncan (1978) found a similar relationship between extractable Ca and pH, but not between Mg and pH. Any K lost from colliery spoil by acid leaching is slowly replenished by weathering of the shale. Some Ca and Mg may also be released in this way. Inputs of all three elements are routinely made by fertilizer and lime additions. It has generally been found that there is very little or no response by plants to added K (Doubleday, 1972; Fitter & Bradshaw, 1974; Pulford *et al.*, 1988).

Marrs & Bradshaw (1980) and Marrs *et al.* (1980) measured very large losses of K from china clay wastes. Despite having a large total K content, the waste was of such a coarse, sandy texture that

Table 7.10 Extractable Ca, Mg and K in substrates

Type of waste and method of extraction	Ca (mg kg^{-1})	Mg (mg kg^{-1})	K (mg kg^{-1})
Coal-mine waste			
Ammonium acetate (Fitter & Bradshaw, 1974)	332–1544	10–144	43–86
Ammonium acetate (Palmer, 1978)	56–209	4.8–49	20–700
Acetic acid (Pulford & Duncan, 1978)			
Mean, unburnt waste	4260	276	152
Mean, burnt waste	7880	288	219
China clay waste			
Ammonium acetate (Bradshaw *et al.*, 1975)	85–115	14–28	8–13
Oil shale waste			
Available levels quoted by Bradshaw & Chadwick, 1980	528–630	19–144	27–360

leaching losses were rapid. Marrs *et al.* (1980) found net accumulation rates for K of −662 to +249 (mean −6.3) kg ha^{-1} year^{-1} in the mica dam walls, and −1620 to +996 (mean 91) kg ha^{-1} year^{-1} in sand tips. An overall nutrient budget for reclaimed china clay waste showed a balance of −37 kg K ha^{-1} year^{-1} representing a loss of 195% of added K by leaching (Marrs & Bradshaw, 1980). Most of the K (90% +) in the china clay ecosystems was found in the soil component (Roberts *et al.*, 1980). The behaviour of Ca and Mg in china clay waste was similar to, but not as extreme as, that of K.

Few studies have been made on sulphur in urban soils and substrates from the point of view of its role as a plant nutrient. Much more attention has been given to the input of S from industrial sources, and the resulting acidity and toxicity. Sulphate ions can be held by chemisorption in the same way as phosphate, although usually less strongly. Sulphur is a constituent of soil organic matter, and is mineralized along with C, N and P. Pyrite, FeS_2, is a component of some colliery spoils, and on oxidation produces sulphates. In some cases solid phase sulphates, such as gypsum, $CaSO_4$, or jarosite, $KFe_3(OH)_6(SO_4)_2$, may be found. In view of the occurrence of S in some substrates and the relatively high inputs from atmospheric sources in urban areas, it is unlikely that sulphur deficiency will be a problem in urban soils and substrates.

Micronutrients

Again, relatively little attention has been paid to the role of micronutrients in plant nutrition in urban soils and substrates. Because of the possibility of high inputs of these elements into urban soils, due to industrial contamination or the addition of high metal wastes, their toxic effects have been more closely studied. There is a vast amount of literature regarding this aspect (e.g. Lepp, 1981; Culbard *et al.*, 1988; and Thornton, this volume).

Of the elements defined as micronutrients, Fe, Mn, Cu and Zn occur in soils as cations, while B and Mo are found as anions. The cationic elements can be held by ion exchange, but this is usually a minor process as the exchange sites are occupied by macro-nutrient cations which are present in much greater amounts. The important processes for holding Fe, Mn, Cu and Zn are chemisorption, chelation and precipitation. Boron and molybdenum are thought to be held primarily by chemisorption. Some wastes derived from quarrying or mining may contain material which has one or more of these elements as a major component (e.g. mine tailings), where solubility will be the main factor controlling release of the element.

Palmer (1978) measured EDTA-extractable Cu and Zn in colliery spoils from Yorkshire and found a range of values for both elements of $0-24 \, mg \, kg^{-1}$. Pulford *et al.* (1982) measured the micronutrient contents ($0.5M$ acetic acid extractable) in unburnt colliery spoil at seven sites in central Scotland. They found that the amounts were highly variable, the mean values being $94.8 \, mg \, Fe \, kg^{-1}$, $45.1 \, mg \, Mn \, kg^{-1}$, $4.08 \, mg \, Cu \, kg^{-1}$ and $12.2 \, mg \, Zn \, kg^{-1}$. The main factors controlling extractable Cu and Zn (and Co and Ni to an even greater extent) were the pH and the manganese oxides. Despite the small amounts of manganese oxides present, they had a much greater influence on the adsorption of Cu and Zn than did the iron or aluminium oxides, which

were present in greater amounts. It was suggested that the conditions of fluctuating pH and redox potential found on many colliery spoil sites led to the formation of highly active forms of manganese oxides, which in the absence of other holding mechanisms, provided a control on the solubility of some micronutrients. Pulford *et al.* (1982) found significant positive relationships between pH and acetic acid extractable Mn, Cu, Zn, Ni and Co on seven colliery spoil tips in central Scotland. This effect was particularly found at $pH > 4$ and supports the finding of Kimber *et al.* (1978) that plants of rose bay willow herb growing on colliery spoil of $pH > 4$ had greater contents of Mn and Zn than plants growing at $pH < 4$. Similar results have been obtained by Cornwell & Stone (1973), in coal spoil from Pennsylvania. It may be that these forms of manganese oxide are sufficiently labile to supply Mn and associated trace elements to plants.

Conclusions

Urban soils constitute a wide range of materials, including true soils, degraded soils, contaminated soils and a variety of substrates derived from industrial processes. Very often it is assumed that the provision and cycling of nutrients in these soils will be similar to the same processes in rural and agricultural soils. Little attention has been paid to the effects of any special features found in urban soils. Much more attention has been paid to specific substrates, usually because they can occupy significant areas of land, which are often reclaimed.

Much of the preceding discussion has drawn upon quantitative data from studies on materials which are not soils, but which act as soils in an urban environment. This highlights the paucity of information on nutrient availability in urban soils. Care should be taken when using data from such sources for predicting behaviour in soils. The assumption that urban soils are nutrient poor is often made, although this is not always the case.

In particular, care should be exercised in the assessment of nutrient sufficiency or deficiency. The only way to determine this unambiguously is to measure the uptake of nutrients by plants growing in a soil or substrate. However, as mentioned above in relation to phosphate, discrepancies can arise between results obtained by field and greenhouse studies. A further problem is the choice of the test plant species, as different species can utilize soil nutrient pools to varying degrees. In the case of urban soils, the relatively small areas occupied by a particular soil or substrate, and the degree of heterogeneity, can also throw doubt on conclusions from this type of experiment. The alternative, use of extractants to remove specific forms of nutrient from soil, may not be much better. Such methods are of little use for soil nitrogen studies due to the influence of the organic nitrogen component. Even for nutrients found predominantly in inorganic forms, extraction procedures may not be appropriate. Care should be taken to match the extracting solution to the soil properties (e.g. acid or alkaline) and to the nutrient pool to be measured (i.e. exchangeable, organically bound, etc.). In the case of some substrates, the control on the supply of certain nutrients may depend on the solubility of one or more component of the substrate. This concept of availability is one which is commonly used for agricultural soils, but even there anomalies and special cases are recognized. For certain soils adjustments are made to the rate of fertilizer applied to allow for over- or underestimation by extraction procedures. Despite these caveats there is a tendency to require a figure for the availability of nutrients to plants. It would therefore be profitable to make an assessment of the suitability of soil extractants for measuring nutrient availability in urban soils, and to attempt to standardize such methods.

A continuing theme throughout this chapter has been the role of biologically-mediated processes of nutrient cycling and supply, which is particularly relevant in the case of nitrogen. Associated with this is the importance of organic matter in the soil, and the need to build up and maintain organic matter contents. Conditions in urban soils may not be conducive to the maintenance of an organic matter component or the viability of biologically mediated reactions. This is an area of the behaviour of urban soils about which much more needs to be known. It is also an important aspect when soils have been moved and stored (Rimmer, this volume).

As may be expected, the greatest attention has been paid to the provision and cycling of nitrogen in urban soils and substrates, as this is often the limiting nutritional factor. Although information is available from a number of sources, studies have tended to focus on certain substrates. There is a need for a better understanding of the controls on the various processes involved in nitrogen cycling in urban soils and substrates. In many cases it would seem that the limiting factor is simply the amount of N present in the system. The current consensus is that legumes are the best way of building up a reservoir of nitrogen. This implies that time will be available for a natural system of cycling to develop. In the future, there may be demands on urban land which would not allow such a slow build up, e.g. increased application of sewage sludge to land in urban areas. Under such a system, large amounts of N would be added to the soil on a regular basis. Nitrogen supply to plants would not be a problem, but more needs to be known about the transformations and losses of N.

Phosphorus supply in urban soils and substrates is generally considered to be next in importance to nitrogen supply. Most attention has been paid to P supply in certain substrates, such as colliery spoil, which tend to be low in total and available P. The use of extractants to assess available P in such materials is uncertain. There is a lack of clear evidence for plant response to added P under field conditions, although a response can usually be demonstrated in the glasshouse. The role of organic P in such systems is unclear. A major requirement is a detailed study of P release by urban soils, and its utilization by plants under field conditions.

Supply of the other macronutrients is not a great problem in most urban soils and substrates. The situation with micronutrients is less clear, as any imbalance is often masked by other factors. In many urban systems an over-supply of micronutrients is more likely due to contamination (Thornton, this volume).

References

Agricultural Development and Advisory Service (1979). *Poultry Manure as a Fertilizer*. Advisory Leaflet 320. HMSO.

Agricultural Development and Advisory Service (1982). *The use of Sewage Sludge on Agricultural Land*. Booklet 2409. HMSO.

Aldag, R.W. & Strzyszcz, Z. (1980) Inorganic and organic nitrogen compounds in carbonaceous phyllosilicates on spoils with regard to forest reclamation. *Reclamation Review* **3**, 69–73.

Allen, S.E., Carlise, A., White E.J. & Evans C.C. (1968).

The plant nutrient content of rainwater. *Journal of Ecology* **56**, 497–504.

Berg, W.A. (1978). Limitations in the use of soil tests on drastically disturbed lands. In Schaller F.W. & Sutton, P. (eds), *Reclamation of Drastically Disturbed Lands*, pp. 653–664. American Society of Agronomy, Madison.

Berg, W.A., Barrau E.M. & Rhodes, L.A. (1975). Plant growth on acid molybdenum mill tailings as influenced by liming, leaching and fertility treatments. In Chadwick M.J. & Goodman G.T. (eds), *The Ecology of Resource Degradation and Renewal*, pp. 207–222. British Ecological Society Symposium No. 15, Blackwell Scientific Publications, Oxford.

Bloomfield, H.E., Handley, J.F. & Bradshaw, A.D. (1982). Nutrient deficiencies and the aftercare of reclaimed derelict land. *Journal of Applied Ecology* **19**, 151–158.

Bradshaw, A.D. (1983). The reconstruction of ecosystems. *Journal of Applied Ecology* **20**, 1–17.

Bradshaw, A.D. & Chadwick, M.J. (1980). *The Restoration of Land*. Studies in Ecology Vol. 6. Blackwell Scientific Publications, Oxford.

Bradshaw, A.D., Dancer, W.S., Handley J.F. & Sheldon, J.C. (1975). The biology of land revegetation and the reclamation of the china clay wastes of Cornwall. In Chadwick, M.J. & Goodman, G.T. (eds), *The Ecology of Resource Degradation and Renewal*, pp. 363–384. British Ecological Society Symposium No. 15, Blackwell Scientific Publications, Oxford.

Bremner, J.M. & Mulvaney, C.S. (1982). Nitrogen – total. In Page, A.L., Miller, R.H. & Keeney, D.R. (eds), *Methods of Soil Analysis*, Part 2, pp. 595–624. Agronomy Monograph 9 (2nd Edition). American Society of Agronomy, Madison.

Cornwell, S.M. & Stone, E.L. (1968). Availability of nitrogen to plants in acid coal mine spoils. *Nature* **217**, 768.

Cornwell, S.M. & Stone, E.L. (1973). Spoil type lithology and foliar composition of *Betula populifolia*. In Hutnik, R.J. & Davis, G. (eds), *Ecology and Reclamation of Devastated Land*, Vol. 1, pp. 105–120. Gordon and Breach, New York.

Culbard, E.B., Thornton, I., Watt, J., Wheatley, M., Moorcroft, S. & Thompson, M. (1988). Metal contamination in British urban dusts and soils. *Journal of Environmental Quality* **17**, 226–234.

Daft, M.J., & Hacskaylo, E. (1976). Arbuscular mycorrhizas in the anthracite and bituminous coal wastes of Pennsylvania. *Journal of Applied Ecology* **13**, 523–531.

Dancer, W.S. (1984). Soil tests for predicting plant available phosphorus in newly reclaimed alkaline minespoil. *Communications in Soil Science and Plant Analysis* **15**, 1335–1350.

Dancer, W.S., Handley, J.F. & Bradshaw, A.D. (1977a). Nitrogen accumulation in kaolin mining wastes in Cornwall. I. Natural communities. *Plant and Soil* **48**, 153–167.

Dancer, W.S., Handley, J.F. & Bradshaw, A.D. (1977b). Nitrogen accumulation in kaolin mining wastes in Cornwall. II. Forage legumes. *Plant and Soil* **48**, 303–314.

Dancer, W.S., Handley, J.F. & Bradshaw, A.D. (1979). Nitrogen accumulation in kaolin mining wastes in Cornwall. III. Nitrogen fertilizers. *Plant and Soil* **51**, 471–484.

Dennington, V.N. & Chadwick, M.J. (1978). The nutrient budget of colliery spoil tip sites. I. Nutrient input in rainfall and nutrient losses in surface run-off. *Journal of Applied Ecology* **15**, 303–316.

Doubleday, G.P. (1972). Development and management of soils on pit heaps. In *Landscape Reclamation*, Vol. 2, pp.25–35. University of Newcastle upon Tyne. IPC Business Press, London.

Doubleday, G.P. & Jones, M.A. (1977). Soils of reclamation. In Hackett, B. (ed.), *Landscape Reclamation Practice*, pp.85–124. IPC Press, Guildford.

Ebelhar, M.W., Barnhisel, R.I., Akin G.W. & Powell, J.L. (1982). Effect of lime, N, P and K amendments to surface-mined coal spoils on yield and chemical composition of common bermudagrass (*Cynodon dactylon*, Kentucky). *Reclamation and Revegetation Research* **1**, 327–336.

Fitter, A.H. (1974). A relationship between phosphorus requirement, the immobilization of added phosphate, and the phosphate buffering capacity of colliery shales. *Journal of Soil Science* **25**, 41–50.

Fitter, A.H. & Bradshaw, A.D. (1974). Responses of *Lolium perenne* and *Agrostis tenuis* to phosphate and other nutritional factors in the reclamation of colliery shale. *Journal of Applied Ecology* **11**, 597–608.

Fitter, A.H., Handley, J.F., Bradshaw, A.D. & Gemmell, R.P. (1974). Site variability in reclamation work. *Landscape Design* **106**, 29.

Frankenberger, W.T. & Tabatabai, M.A. (1980). Amidase activity in soils. I. Method of assay. *Soil Science Society of America Journal* **44**, 282–287.

Frankenberger, W.T. & Tabatabai, M.A. (1982). Transformations of amide nitrogen in soils. *Soil Science Society of America Journal* **46**, 280–284.

Fresquez, P.R. & Lindemann, W.C. (1983). Greenhouse and laboratory evaluations of amended coal-mine spoils. *Reclamation and Revegetation Research* **2**, 205–215.

Fresquez, P.R., Aldon, E.F. & Lindemann, W.C. (1987). Enzyme activities in reclaimed coal mine spoils and soils. *Landscape and Urban Planning* **14**, 359–367.

Gemmell, R.P. & Goodman, G.T. (1980). The maintenance of grassland on smelter wastes in the Lower Swansea Valley. III. Zinc smelter waste. *Journal of Applied Ecology* **17**, 461–468.

Jefferies, R.A., Bradshaw, A.D. & Putwain, P.D. (1981). Growth, nitrogen accumulation and nitrogen transfer by legume species established on mine spoils. *Journal of Applied Ecology* **18**, 945–956.

Jenkinson, D.S. & Powlson, D.S. (1976). The effect of biocidal treatments on metabolism in soil. V. A method for measuring soil biomass. *Soil Biology Biochemistry* **8**, 209–213.

Keeney, D.R. & Nelson, D.W. (1982). Nitrogen – inorganic forms. In Page, A.L., Miller, R.H. & Keeney, D.R. (eds), *Methods of Soil Analysis*, Part 2, pp. 643–698. Agronomy Monograph 9 (2nd Edition). American Society of Agronomy, Madison.

Khan, M.Q. (1987). *Studies on the Measurements of Extractable and Mineralizable Nitrogen in Soil*. PhD Thesis, University of Glasgow.

Killham, K. & Wainwright, M. (1981). Microbial release of sulphur ions from atmospheric pollution deposits. *Journal of Applied Ecology* **18**, 889–896.

Kimber, A.J., Pulford, I.D. & Duncan, H.J. (1978). Chemical variation and vegetation distribution on a coal waste tip. *Journal of Applied Ecology* **15**, 627–633.

Lepp, N.W. (ed.) (1981). *Effect of Heavy Metal Pollution on Plants*. Vol. 1. *Effects of Trace Metals on Plant Function*. Vol. 2, *Metals in the Environment*. Applied Science Publishers, London.

Lister, J.E. (1987). *The Nutrient Status of Colliery Spoil in Central Scotland*. PhD Thesis, University of Glasgow.

McGill, W.B. (1977). Soil restoration following oil spills – a review. *Journal of Canadian Petroleum Technology* **16**, 60–67.

Marrs, R.H. & Bradshaw, A.D. (1980). Ecosystem development on reclaimed china clay wastes. III. Leaching of nutrients. *Journal of Applied Ecology* **17**, 727–736.

Marrs, R.H., Roberts, R.D. & Bradshaw, A.D. (1980). Ecosystem development on reclaimed china clay wastes. I. Assessment of vegetation and capture of nutrients. *Journal of Applied Ecology* **17**, 709–717.

Olesen, S.E., Ovig, J.K. & Grant, R.O. (1984). The effects of sewage sludge and lime on vegetation of abandoned brown coal pits. *Reclamation and Revegetation Research* **2**, 267–277.

Olsen, S.R. & Sommers, L.E. (1982). Phosphorus. In Page, A.L., Miller, R.H. & Keeney, D.R. (eds), *Methods of Soil Analysis*, Part 2, pp. 403–430. Agronomy Monograph 9 (2nd Edition). American Society of Agronomy, Madison.

Otrosina, W.J., Marx, D.H. & May, J.T. (1984). Soil microorganism populations on kaolin spoil with different vegetative covers. *Reclamation and Revegetation Research* **3**, 1–15.

Ozanne, P.G. & Shaw, T.C. (1967). Phosphate sorption by soils as a measure of the phosphate requirement for pasture growth. *Australian Journal of Agricultural Science* **18**, 601–612.

Palmer, J.P. (1987). The effect of nitrogen level and time of sowing on the establishment of white clover (*Trifolium repens*) on colliery soil. *Landscape and Urban Planning* **14**, 369–378.

Palmer, J.P. & Chadwick, M.J. (1985). Factors affecting the accumulation of nitrogen in colliery spoil. *Journal of Applied Ecology* **22**, 249–257.

Palmer, J.P. & Iverson, L.R. (1983). Factors affecting nitrogen fixation by white clover (*Trifolium repens*) on colliery spoil. *Journal of Applied Ecology* **20**, 287–301.

Palmer, J.P., Morgan, A.L. & Williams, P.J. (1985). Determination of the nitrogen composition of colliery spoil. *Journal of Soil Science* **36**, 209–217.

Palmer, J.P., Williams, P.J., Chadwick, M.J., Morgan, A.L. & Elias, C.O. (1986). Investigations into nitrogen sources and supply in reclaimed colliery spoil. *Plant and Soil* **91**, 181–194.

Palmer, M.E. (1978). Acidity and nutrient availability in colliery spoil. In Goodman, G.T. & Chadwick, M.J. (eds), *Environment Management of Mineral Wastes*, pp. 85–126, Sijthoff & Noordhoff, Netherlands.

Power, J.F., Bond, J.J., Sandoval, F.M. & Willis, W.O. (1974). Nitrification in Paleocene shale. *Science* **183**, 1077–1079.

Pulford, I.D. (1976). *Reclamation of Industrial Waste and Derelict Land*. PhD Thesis, University of Glasgrow.

Pulford, I.D. & Duncan, H.J. (1975). The influence of pyrite oxidation products on the adsorption of phosphate by coal mine waste. *Journal of Soil Science* **26**, 74–80.

Pulford, I.D. & Duncan, H.J. (1978). The influence of acid leaching and ignition on the availability of nutrients in coal mine waste. *Reclamation Review* **1**, 55–59.

Pulford, I.D., Kimber, A.J. & Duncan, H.J. (1982). Influence of pH and manganese oxides on the extraction and adsorption of trace metals in colliery spoil from the central Scotland coalfield. *Reclamation and Revegetation Research* **1**, 19–31.

Pulford, I.D., Flowers, T.H., Shah S.H. & Walker, T.A.B. (1988). Supply and turnover of N, P and K in reclaimed coal mine waste in Scotland. In *Proceedings of the Mine Drainage and Surface Mine Reclamation Conference*, Pittsburgh, April 1988, Vol. II. Mine Reclamation, Abandoned Mine Lands and Policy Issues, pp. 228–235. US Bureau of Mines, Information Circular 9184.

Reeder, J.D. & Berg, W.A. (1977). Nitrogen mineralization and nitrification in a Cretaceous shale and coal mine spoils. *Soil Science Society of America Journal* **41**, 922–927.

Roberts, R.D., Marrs, R.H. & Bradshaw, A.D. (1980).

Ecosystem development on reclaimed china clay wastes. II. Nutrient compartmentation and nitrogen mineralization. *Journal of Applied Ecology* **17**, 719–725.

Russell, E.W. (1973). *Soil Conditions and Plant Growth*, (10th Edition). Longmans, London.

Shah, S.S.H. (1988). *Transformations of Nitrogen and its Availability to Plants in Coal Mine Soils*. PhD Thesis, University of Glasgow.

Skeffington, R.A. & Bradshaw, A.D, (1980). Nitrogen fixation by plants grown on reclaimed china clay waste. *Journal of Applied Ecology* **17**, 469–477.

Skeffington, R.A. & Bradshaw, A.D. (1982). Nitrogen accumulation in kaolin mining wastes in Cornwall. IV Sward quality and the development of a nitrogen cycle. *Plant and Soil* **62**, 439–451.

Sopper, W.E. & Seaker, E.M. (1983). *A Guide for Revegetation of Mined Land in Eastern United States Using Municipal Sludge*. University Park: The Pennsylvania State University, Institute for Research on Land and Water Resources.

Stevenson, F.J. (1982a). *Humus Chemistry: Genesis, Composition and Reactions*. John Wiley and Sons, New York.

Stevenson, F.J. (1982b). Nitrogen – organic forms. In Page, A.L., Miller, R.H. & Keeney, D.R. (eds), *Methods of Soil Analysis*, Part 2, pp. 625–641. Agronomy Monograph 9 (2nd Edition). American Society of Agronomy, Madison.

Stevenson, F.J. (ed.) (1982c). *Nitrogen in Agricultural Soils*. Agronomy Monograph 22. American Society of Agronomy, Madison.

Stroo, H.F. & Jencks, E.M. (1982). Enzyme activity and respiration in minesoils. *Soil Science Society of America Journal* **46**, 548–553.

Tabatabai, M.A. (1977). Effects of trace elements on urease activity in soils. *Soil Biology Biochemistry* **9**, 9–13.

Tabatabai, M.A. & Alkhafaji, A.A. (1980). Comparison of nitrogen and sulfur mineralization in soils. *Soil Science Society of American Journal* **4**, 1000–1006.

Tabatabai, M.A. & Bremner, J.M. (1972). Assay of urease activity in soils. *Soil Biology Biochemistry* **4**, 479–487.

Tabatabai, M.A., Burwell, R.E., Ellis, B.G., Keeney, D.R., Logan, T.J., Nelson, D.W., Olson, R.A., Randall, G.W., Timmons, D.R., Verry, E.S. & White, E.M. (1981). *Nutrient Concentrations and Accumulations in Precipitation Over the North Central Region*. Research Bulletin 594, Agriculture and Home Economics Experiment Station, Iowa State University of Science and Technology, Ames, Iowa.

Topper, K.F. & Sabey, B.R. (1986). Sewage sludge as a coal mine spoil amendment for revegetation in Colorado. *Journal of Environmental Quality* **15**, 44–49.

Townsend, W.N. & Gillham, E.W.F. (1975). Pulverized fuel ash as a medium for plant growth. In Chadwick M.J. & Goodman G.T. (eds), *The Ecology of Resource Degradation and Renewal*, pp. 287–304. British Ecological Society Symposium No. 15, Blackwell Scientific Publications, Oxford.

Voos, G. & Sabey, B.R. (1987). Nitrogen mineralization in sewage sludge amended coal mine spoil and topsoils. *Journal of Environmental Quality* **16**, 231–237.

White, R.E. (1979). *An Introduction to the Principles and Practice of Soil Science*. Blackwell Scientific, Oxford.

White, R.E. (1980). Retention and release of phosphate by soil and soil constituents. In P.B. Tinker (ed.) *Soils and Agriculture*, pp. 71–114. Blackwell Scientific Publications, Oxford.

Wild, A. (ed) (1988). *Russell's Soil Conditions and Plant Growth*. 11th Edition. Longman Scientific and Technical, Harlow.

Williams, P.J. & Cooper, J.E. (1976). Nitrogen mineralization and nitrification in amended colliery spoil. *Journal of Applied Ecology* **13**, 533–543.

Young, J.L. & Aldag, R.W. (1982). Inorganic forms of nitrogen. In Stevenson, F.J. (ed) *Nitrogen in Agricultural Soils*. pp. 43–66. Agronomy Monograph 22, American Society of Agronomy, Madison.

Zantua, M.I. & Bremner, J.M. (1975). Comparison of methods for assaying urease in soils. *Soil Biology Biochemistry* **7**, 291–295.

8 The biology of soils in urban areas

J.A. HARRIS

Introduction: the soil biota

The living components of soil, the biota, are fundamental to the development and maintenance of the soil ecosystem. They play important roles in incorporation of organic matter, decomposition, mineralization and nutrient cycling, and in the development and maintenance of soil structure. The focus of research into the biology of soils has been on natural and agricultural systems, with scant attention being paid to those in the urban environment. In recent years, however, there has been much interest in the effects on substrates either disturbed or made available by industrial activity, such as opencast coal mining. It is to this area of research that we must look for some of our evidence.

Soil organisms may be classified into three groups, on the basis of size (Table 8.1), as defined by Richards (1987): (i) the *microbiota*, including the bacteria, fungi, algae, protozoa and viruses; (ii) the *mesobiota*, including smaller arthropods and enchytraeids, nematodes and springtails, and (iii) the *macrobiota*, including earthworms, molluscs, and the larger arthropods and enchytraeids.

Some authors regard plant roots as components of the macrobiota, but they are not truly soil organisms, although they are important contributors to the soil ecosystem. Similarly, burrowing animals and insects are only passing visitors compared to the organisms listed above. The macro- and meso-biota can be considered as one group, the soil fauna, and this term will be used in this chapter. It is beyond the scope of this chapter to give a detailed description of the groups, although a generalized description will provide some useful background information.

THE SOIL FAUNA

There is a great assortment of animals to be found in the soil; some animals spend their entire lives there and others are only present for part of their life cycles but are no less important. Earthworms, nematodes, some mites and springtails are amongst the former group. Animals which are only present as resting stages, e.g. as eggs, cysts or pupae, are usually excluded from this general group, although they may provide food for some soil inhabitants.

There are several ways in which the fauna may be further classified including habitat occupied, method of locomotion, method of feeding, and the size or duration of time spent in the soil. No single criterion is completely satisfactory but used together they may be applied with some success.

The microfauna, which include all of the protozoa and some mites, rotifers and nematodes, is a subdivision of the microbiota and will be considered later. Suffice to say at this stage that they play little part in comminution of litter and therefore form a coherent group (Swift *et al.*, 1979).

The numbers of these animals found in similar soil systems vary greatly; the ranges in temperate grassland ecosystems are given in Table 8.2, and their major ecosystem functions in Table 8.3.

THE MICROBIOTA

The major division within the microbiota is between the eukaryotes and prokaryotes. The eukaryotes are represented by the algae, protozoa and fungi. The soil prokaryotes are bacteria and actinomycetes. Although viruses are found in the soil their functional significance is unclear at this stage, save as pathogens of a number of species.

The algae are photosynthetic organisms, which are predominantly aquatic, but some are present in moist or wet terrestrial systems. They are frequently unicellular, may be present in large numbers in the soil surface, and may sometimes be found at depth. Free-living protozoa are abundant and are important predators of bacteria.

The fungi are commonly associated with soil, and by ramifying into dead material begin the processes of decay and nutrient recycling. The wood-decaying fungi are responsible for returning carbon to the soil in forest systems. Mycorrhizal fungi form important symbiotic associations with plant roots; the fungus

Table 8.1 Size classification of soil organisms

Class	Size (mm)
Microbiota	< 0.2
Mesobiota	0.2–10.4
Macrobiota	> 10.4

Table 8.2 Numbers of major soil animals

Group	Thousands per square metre
Nematodes	$1.8 \times 10^3 – 120 \times 10^3$
Mites	20–120
Collembola	10–40
Enchytraeid worms	0.5–20
Molluscs	0.5–9
Earthworms	0.1–2
Larger myriapods	0.5–2
Beetles and larvae	0.5–1
Dipterous larvae	0.5–0.9
Spiders	0.2–0.9
Ants	0.2–0.5
Isopods	0.2–0.4

gains photosynthate from the plant in return for access to nutrient pools that would normally be unavailable to the plant root. This association is extremely sensitive to environmental stress and disturbance (Read & Birch, 1988), which has major implications for revegetation of soils found in urban areas.

Fungi are also implicated in the formation of microaggregates, by binding together soil mineral and organic particles with hyphae. They are also responsible for the majority of plant diseases, and this will be of particular importance when plants are under stress in soils which may not be ideal substrates.

The slime moulds (myxomycota) are fungus-like organisms. During part of their life cycle they exist as amoeboid forms, but come together in spore-

Table 8.3 Major soil animals: distribution, feeding behaviour and ecosystem functions

Group	Distribution in profile layers	Range of feeding strategies	Involvement in organic matter decomposition	Involvement in soil structure formation
Nematodes	Litter, humus & mineral	Herbivores & carnivores	No	No
Rotifers	Litter & humus	Omnivores, herbivores & carnivores	Limited	No
Enchytraeids	Litter, humus & mineral	Herbivores & carnivores	No	Form channels
Earthworms	Humus & mineral	Detritivores	Yes	Very important
Snails & slugs	Clumped, in litter	Detritivores & herbivores	Limited	No
Arthropods	Litter, humus & mineral	Detritivores & carnivores	Yes	Yes, forming channels
Isopods	Litter, humus & mineral	Detritivores	Yes	No
Myriapods	Litter & humus	Detritivores & Herbivores	Yes	No
Insects	Litter, humus & mineral	Omnivores	Yes	Form channels and pores

forming bodies. They prey on bacteria and may have a significant role in forest ecosystems.

The bacteria mediate many soil nutrient cycles, for example nitrogen mineralization, nitrification and denitrification. They are ubiquitous in soil systems and no soil can function without them to transform nutrients. They are an extremely adaptable group found in anaerobic, hot, saline and cold environments. They also, like the fungi, form a labile nutrient pool.

Actinomycetes are economically important because they are a source of antibiotics such as streptomycin. They form filamentous structures within the soil system and can compete with bacteria by chemical antagonism.

The cyanobacteria, formerly known as blue-green algae, are photosynthetic organisms. They are unlikely to provide a major source of fixed carbon in systems where there is any sort of vegetation cover.

FACTORS AFFECTING COMMUNITY STRUCTURE

The size and composition of the plant community are largely determined by two factors, stress and disturbance (Grime, 1979). Stress may be defined as a factor which limits the growth of the majority of the organisms within the community, such as nutrient limitation or the presence of toxic substances. This will, in effect, define the 'carrying capacity' or total biomass capable of being sustained by the habitat. Disturbance is an event which brings about the sudden availability of substrates for growth. This may be an enrichment disturbance, such as litter fall, or a destructive disturbance such as earthmoving or fire. It is clear that these factors are likely to be operating on the soil system, with many features of stress (e.g. pollution) or disturbance (e.g. housing construction, trampling), occurring in urban areas.

SOIL NUTRIENT CYCLING

Soil organisms are responsible for incorporation of organic matter into the soil and cycling of nutrients. In some instances certain groups are solely responsible for these changes, e.g. the conversion of ammonium to nitrate via nitrite (nitrification) is carried out by a restricted group of bacteria. Higher soil organisms are responsible for initial fragmentation of plant litter when it first falls, or when plant roots die, and as the pieces of organic debris get smaller and more accessible, so other groups take over, such as fungi which ramify into the substrate.

SOIL STRUCTURE FORMATION AND MAINTENANCE

Soil organisms have a major role to play in the development and maintenance of a stable soil structure, alongside plant roots. Animals such as earthworms form channels as a result of their burrowing activities, and they excrete pellets which contribute to the formation of stable aggregates. Fungi contribute to microaggregate formation by ramifying through the soil and binding particles together by enmeshment in their hyphae. Bacteria secrete polysaccharides which stick particles together.

One important feature of soil nutrient cycling and structure-forming processes is that they are temporary or transient. The structure must be maintained by a constant turnover of polysaccharides, hyphae, plant roots and their exudates. If this essentially dynamic process is halted in any way then the structure is likely to deteriorate. Similarly, certain nutrient cycles may be affected by stress or disturbance. These two factors will affect the potential size and composition of the soil community. This is the case in soils disturbed by man's activities, such as opencast mining, or metal extraction, and will almost certainly be the case in an urban environment where constant pressure from human activity is a feature of the soils. This activity may be as seemingly innocuous as leisure pursuits, but nevertheless the resulting trampling may have deleterious effects on soil structure or nutrient cycling.

The environment of urban soils

In considering the likely consequences of man's activities on the biology of soils in urban areas we can begin by assessing the characteristics of urban soils. Craul (1985) summarized these as follows.
1 Great spatial variability.
2 Compaction leading to modified structure.
3 Presence of a surface crust on bare soil that is usually water repellent.
4 Modified pH.
5 Restricted aeration and drainage of water.

6 Interrupted nutrient cycling and modified activity of soil organisms.

7 Presence of manufactured materials and other contaminants.

8 Modified temperature regimes.

Little research has been done specifically on soils in urban areas but in the light of these characteristics, it is possible to extrapolate from the existing literature on the biology of systems disturbed by industrial activity. In particular, in recent years there has been information gained from work on opencast and deep mines, and areas contaminated as a result of metal ore workings. Areas worked by the opencast method tend to contain soils which are highly compacted as a result of the use of heavy earthmoving equipment, and have the features **1, 2, 5, 6** and **7** listed above. Sites with metal contamination indicate the likely situation in similarly affected urban soils, and have features **7** and **8**. Where information exists on work done in urban areas this has been given precedence.

The research carried out has been summarized on the basis of several themes, in an attempt to provide some insight into the biology of soils in urban areas. This includes evidence from work on the effects of disturbance, on 'spoil' which is in effect any one of a number of 'soil-forming' materials which may be used in modern housing schemes as a base substrate for providing garden areas, implications of soil compaction, implications of pollution, work done to encourage soil organisms which could be applied to urban areas, and finally reports of that work which has been carried out in urban areas.

THE EFFECTS OF DISTURBANCE ON SOIL ORGANISMS

In 1982, Standen *et al.* reported studies on earthworm populations carried out on areas restored from opencast mines and spoil from deep mines. They noted that lumbricid populations were small compared to permanent pasture areas nearby, even in areas 11 years after restoration. They recommended that certain species be reintroduced into soils after restoration, or that mechanical disturbance during the restoration period be reduced by employing direct drilling of winter forage crops, instead of ploughing and reseeding for hay.

Rushton (1986a) investigated the consequences

of opencast mining on earthworm populations in Northumberland. He recorded small populations in newly reclaimed areas but, in contrast to Standen *et al.* (1982), found larger numbers in some 15-year-old sites. He attributed this difference to the greater amounts of manurial application being made at the sites he studied. There were, however, differences in community structure on the five reclaimed sites, as compared to the undisturbed controls. *Lumbricus terrestris* was absent in all but one of the reclaimed sites. This may have been due to a combination of low food availability and continued disturbance by ploughing, although the presence of *Aporrectodea longa* does not fully support this hypothesis. Also there may have been direct competition between these species. For *L. terrestris* to make an effective contribution to the formation of drainage channels on restored sites, pastures need to be disturbed as little as possible.

Scullion *et al.* (1988) reported that cultivation treatments resulted in decreased numbers of earthworms on both worked and undisturbed controls but that the effects were transitory on the unworked land. This adds further weight to the argument against cultivating restored or newly created sites in favour of surface applied organic manures followed by grazing regimes.

Several workers have reported the accumulation of a mat of undecomposed litter on disturbed sites (Marrs *et al.*, 1980; Schafer *et al.*, 1980; Stroo & Jencks, 1982). These studies suggest that what were formerly mull soils are being converted to mor soils as a result of opencast mining. This may also pertain in urban areas where disturbed soil or soil substitutes are used as a substrate for vegetation cover.

It would appear that disturbance leads to decreases in community size and structure, with certain processes (e.g. litter incorporation) being particularly adversely affected.

BIOLOGY OF SPOIL AND SOIL-FORMING MATERIALS

Vimmerstedt & Finney (1973) investigated the potential to establish earthworms in coal-spoil banks within a short time after revegetation, and measured the effect of earthworm activity on soil formation. They demonstrated that introduced earthworms could survive and reproduce in spoil of pH 3.5–4.0

and persist for at least five years. There was also good evidence that the worms incorporated surface litter into the spoil profile. They concluded that the introduction of earthworms was not only feasible, but recommended it as a means of enhancing species diversity in disturbed areas, with the aim of achieving a speedier return to a functioning, self-sustaining, ecosystem.

Early work by Wilson & Stewart (1955, 1956) indicated that strip-mine spoils had small concentrations of nitrogen but possessed a microbial population capable of organic matter decomposition even in the absence of a vegetative cover. To assess the effect on the decomposition of adding nutrients to spoil a series of laboratory experiments was carried out (Hedrick & Wilson, 1956). They reported the greatest increase in carbon dioxide evolution when nitrogen was added, whether alone or in combination with phosphorus, potassium, straw or calcium hydroxide. Calcium hydroxide addition, to achieve a more favourable pH, was secondary in importance to nitrogen addition, whilst phosphorus and potassium additions had little effect.

In a study of the microbiology of an Appalachian strip-mine spoil, Wilson (1965) found markedly greater numbers of fungi, bacteria and actinomycetes in a revegetated and an undisturbed site when compared to unvegetated spoil. Wilson concluded that vegetation exerted a greater influence on the soil microflora than the low pH caused by sulphur-oxidizing bacteria. The diversity of physiological groups was also greater in the revegetated spoil, i.e. ammonifiers, nitrifiers, nitrate reducers, cellulose degraders and polysaccharide producers.

Schramm (1966) demonstrated a relationship between the colonizing success of plant species in Pennsylvanian coal spoils and the presence of certain ectomycorrhizal fungi and nitrogen-fixing bacteria, work later mirrored by that of Daft & Nicolson (1974) in Scottish coal spoils. Schramm (1966) also noted low bacterial biomass generally in coal-mine spoils.

Miller & Cameron (1978) reported the results of a major investigation into the microflora of two abandoned mines in the mid-western United States. The data revealed that the nature of refuse (i.e. spoil) materials was unfavourable for survival and growth of soil organisms. Despite the establishment of pioneer populations such as acidophilic fungi and some acid-tolerant algae, it was recommended that the refuse materials should be ameliorated to raise the pH and to provide nutrients, especially nitrogen. It was suggested that biodegradable organic matter should also be added for its physical and biological effects. The authors pointed out that although 60 years had passed at one site since it had been abandoned, little colonization apart from at the periphery, had occurred. At this boundary, on the protected northern treeline/refuse interface, mosses, lichens and a diverse microflora had established, which should eventually lead to colonization by macrophytes and cryptogams, and increased nutrient cycling. The emphasis, then, was that there was a requirement to compress this successional process into a time scale where successful reclamation could be witnessed by legislators and the electorate. It is evident that without appropriate treatment these sorts of materials will not develop a self-sustaining ecosystem in reasonable time scales.

Hersman & Klein (1979) in a study of retorted oil shales, found a good correlation between soil dehydrogenase activity and numbers of fungi and bacteria, work confirmed later by Fresquez & Lindemann (1982).

In 1981, Miller & May produced the final report for the Staunton 1 Reclamation Demonstration Project which was a continuation of earlier work (Miller & Cameron, 1978) and investigated reclamation techniques suggested by it. After one to five years none of the reclaimed plots equalled an old field control in decomposition rates or microbial respiration. However, numbers of microorganisms approached those in the control, but had much smaller species diversity. This implies that the composition of the microbial community had become altered, to one incapable of carrying out various stages of organic matter decomposition. This community may be dominated by 'zymogenous' bacteria capable of short-term, rapid growth. It would also lack certain fungi, as shown by the low scores of 0.7–1.5 on Brillouin's Diversity Index for the disturbed areas, as compared to 2.1–2.2 for the control plots.

Stroo & Jencks (1982) sampled a number of strip-mine soils of varying type and age in Preston County, West Virginia, primarily to assess whether there was development of self-sustaining ecosystems on reclaimed strip-mine sites, and to measure differences in the effectiveness of reclamation treatments.

They used a variety of physiological, enzymatic and physico-chemical methods. All the measurements indicated that minesoils had lower microbial activities than undisturbed controls, but that amylase and phosphatase activities increased with time, as did respiration rates, to reach similar levels to those in the control site 20 years after restoration. This recovery in microbial activity was linked to accumulation of carbon and nitrogen. However, on sites which had been fertilized and limed they found very large respiration rates, with low to moderate amounts of organic matter and nitrogen. There were also decreases in microbial activity and carbon and mineralizable nitrogen with age in the areas amended with lime and fertilizer, which suggested rapid decomposition and erosion losses. They concluded that the readily decomposable organic matter is attacked by a zymogenous population, when inorganic nitrogen is supplied as a fertilizer. This leads to an increasing proportion of resistant or recalcitrant organic matter being left, which could in turn lead to decreased productivity and stability against erosion once the application of fertilizer is discontinued. One site, which had been subject to such a strategy for some 18 years, served as a good example of this; the vegetation had failed after ten years and had become barren, with activity levels similar to an adjacent non-revegetated site. They also pointed out that initial revegetation should not be the only criterion of successful reclamation. They concluded that vegetation was critical to the recovery of microbial activity, with nitrogen fixation by legumes being of paramount importance. The problems presented by compaction (which slowed recovery) were underlined and low phosphatase activity indicated possible future problems with supply of available phosphorus. They end by stating that 'the loss of nitrogen and organic matter in young, amended minesoils suggests that the goal of recreating productive, self-sustaining ecosystems is not being achieved on such sites and serious problems of declining productiviy and increasing erodability may result from current reclamation practices'. So even with the research done to date in the United States the problem of 'old-method carryover' remained.

Visser *et al.* (1983) investigated a prairie site in Alberta, Canada. A number of microbial factors were found to be significantly lower in the mined soil as compared to the control. These were soil respiration, microbial biomass C, ATP, actinomycete numbers, hyphal lengths and nitrogen-fixing potential. Bacterial numbers were, however, greater in the disturbed soil and dominated by coryneforms, whereas in the undisturbed control *Bacillus* spp. and non-pigmented Gram-negative rods were also well represented. There was also a shift in the composition of the fungal species from one dominated by *Chrysosporium – Pseudogymnoascus* spp. and sterile dark forms, to one dominated by *Alternaria* spp., *Clasdosporium* spp., sterile dark forms and yeasts, although little functional significance was attached to this shift by the authors. There were more *Penicillium* spp. occurring in the disturbed soils, and this difference was ascribed to them being among the most common airborne saprophytic fungi, being able to grow on a wide variety of substrates and also able to withstand a wide variety of environmental extremes, including low water potentials and high temperatures. Visser and co-workers also found greater rates of decomposition of cellulose filter papers placed on the soil surface in nylon mesh bags in the disturbed site as compared to the undisturbed control. They ascribed this to differing micro-environmental conditions around the filter papers; they tended to become embedded in the disturbed site. Although the authors called into question the link between decomposition losses of cellulose paper and natural plant litter, this difference between the two sites does support the argument forwarded by Stroo & Jencks (1982) that there is potential, given the right conditions, for rapid loss of carbon from disturbed soils. Supplying them with fertilizer-nitrogen could be the wrong thing to do if long-term sustainability is the objective. This did not prevent Visser (1985) recommending fertilizer application to encourage plant growth as a possible reclamation treatment, despite Stroo & Jencks's previous appeals to the contrary. It must be said, however, that this recommendation was accompanied by one for use of organic amendments. It is likely that organic amendments will be required in disturbed or stressed urban soils, as repeated application of inorganic nutrients is unlikely to be sustained.

Visser (1985) supplemented her previous work on prairies with data from sub-alpine and oil sand sites. Both disturbed and undisturbed soils were sampled. In all three cases there were far greater amounts of total fungal mycelium and mycelium with cell contents in the undisturbed as compared to the control site. Also, unlike the prairie soil, the two additional sites had greater numbers of bacteria in the undisturbed controls. All three sites showed delayed CO_2 efflux, and increases in biomass carbon and ATP content on the disturbed sites as compared to the controls. Visser found that woody substrates were unsuitable as indicators of decomposition potential due to slow decay rates, even when amended with ammonium nitrate to reduce the carbon:nitrogen ratio, and she preferred to use the rate decomposition of cellulose filter paper in nylon mesh bags. Faster decay rates were also recorded in the disturbed prairie and subalpine sites, in comparison with their controls. Here the relationship between pure cellulose degradation and plant litter was also called into question, but interestingly the conclusion was reached that because of its better nutritional status, plant litter may decompose even more rapidly than pure substrates.

Fresquez *et al.* (1987) studied the effects of reclamation age, stockpiling of soil and topsoiling versus bare spoil on a number of enzyme activities. They found lower activities in non-topsoiled areas, compared to topsoiled, with enzyme activities recovering in these topsoiled areas and peaking one to two years after reclamation. They concluded that topsoil was extremely useful in re-establishing soil processes after mining.

The importance of the microbial community in establishing a functioning ecosystem in 'soil-forming' materials has been made quite clear in this work. Also the importance of earthworms as initial fragmentors is apparent.

BIOLOGICAL IMPLICATIONS OF SOIL COMPACTION

When soils are brought onto site for incorporation into housing schemes, or are subject to trampling, it is likely that compaction of the surface layers will occur. Little information exists of studies in urban areas, but there are some examples from mineral workings.

Armstrong & Bragg (1984) investigated some soil physical parameters and earthworm populations on opencast sites restored up to 35 years previously. Despite differences in pH they found no pH effects on earthworm recolonization, which tended to re-

turn to the same biomass per unit area as undisturbed sites within 1–20 years of restoration. On restored sites they noted some zones of compaction which prevented earthworms from penetrating deep into the soil profile and resulted in shallow-living earthworms being favoured.

Rushton (1986b) studied the effects of soil compaction on the tunnelling behaviour of *L. terrestris*. He found that increasing bulk density led to a decrease in tunnelling activity and that the deep cultivation required to relieve this condition would prevent the re-establishment of earthworm populations.

White & McDonnell (1988) compared nitrogen cycling in urban versus rural, forest soils. The urban forest was part of the New York Botanical Garden, situated in the Bronx, New York City, USA, and the rural forest was part of the Mary Flagler Cary Arboretum, 117 km north of New York City. Although they did not investigate invertebrates quantitatively, they noticed a lack of earthworms in the soils taken from the urban forest and suggested that trampling by humans had reduced the numbers of all invertebrates. They further suggested that this was part of the reason for the comparatively low nitrogen mineralization in the urban soil. Smeltzer *et al.* (1986) reported this effect in other forest systems.

BIOLOGY OF POLLUTED SOILS

There exists some literature on the effects of metal pollution on vegetation but very little on the effects on soil organisms *in situ*.

Hunter *et al.* (1987) investigated the invertebrate populations of contaminated and semi-contaminated grasslands near a major copper refinery and copper/cadmium alloying plant, and an uncontaminated control. All invertebrates showed marked increases in their total body copper and cadmium concentrations when compared to the control populations. The populations of oligochaetes and isopods were significantly reduced at the refinary site. The decrease in earthworm numbers supports evidence found elsewhere (Neuhauser *et al.*, 1985). Investigations of the transfer of these metals through the food chain revealed that both detritivores and herbivores showed diet concentration factors of two to four times that for copper. However, the results for cadmium were somewhat different with the detritivores showing a concentration factor of ten to twenty times, and the herbivores three to five times. There was a suggestion that there may be a degree of homeostatic control over the accumulation of copper, but not of cadmium. The ratios of copper: cadmium in two carnivore groups, predatory beetles and spiders were also different. The spiders had a copper: cadmium ratio of 9–11:1, as compared to 30–35:1 for the beetles. This work has implications for transfer higher up the food chain as arachnid numbers were not depressed but had the highest cadmium concentrations; therefore predation of spiders by small mammals and birds could lead to accumulation in this biomass pool. This in turn could lead to accumulation in higher predators. The authors also found that there were marked seasonal variations in the metal contents of the invertebrate biomass brought about by variation in the abundance, species composition and age structure of populations. This led to the recommendation that 'single visit' sampling programmes be avoided as the data could not be extrapolated.

Ohya *et al.* (1988) reported on a study of soils from Sakai City, Osaka, part of the Hanshin Industrial Megalopolis, Japan, that investigated the impact of heavy metal pollution on the size and activity of the soil microbial biomass. Thirty soil samples were collected from near roadsides, arable lands and city parks within Sakai. The following physico-chemical analyses were carried out: pH, cation-exchange capacity (CEC), total organic-C, total-N, and heavy metals soluble in dilute acid. Activity of the microbial community was determined by measurement of carbon dioxide evolution from soils incubated with or without added glucose and ammonium sulphate. The amounts of Zn and Pb were found to be ten and nine times higher than the national average, respectively. The values they obtained for the microbial characteristics are given in Table 8.4. Simple regression analysis indicated that bacterial numbers were not significantly related

to any of the physico-chemical characteristics, whereas fungal numbers were correlated to organic-C content. Biomass C was positively correlated to CEC, organic C and total N, a finding consistent with studies of natural systems. The only significant correlation with extractable Zn + Pb was with carbon dioxide evolution. Basal respiration was correlated to biomass C, as would be expected. Multiple regression analysis revealed that the reduction in carbon dioxide evolution in soils incubated with added substrates was explained by easily soluble Zn + Pb, and that the evolution was not closely related to biomass C. They concluded that the dormant biomass was not greatly affected by heavy metals, in contrast to the metabolically active part. This is probably due to little uptake of ions occurring in the dormant biomass. This may help non-tolerant biomass survive heavymetal pollution, but this is of little use in establishing a functioning ecosystem.

Golovacheva *et al.* (1986) examined the corrosion zones around steel heating pipes in the city of Moscow, USSR. A sulphate-reducing bacterium *Desulfotomaculum nigrificans* and *Sulfobacillus thermosulfidooxidans* were found in the heated soil surrounding the corroded pipes. The authors implicated them in the corrosion process, and suggested that a thermally stable niche had been created in these areas. They presented a scheme explaining how these organisms may be involved in steel corrosion.

In a study of coniferous needle decomposition Fritze (1988) found that in forests close to or in urban areas there were some effects of airborne pollutants. The main effects were an increase in the rate of release of Mn from the decomposing needles and enrichment of Fe and Pb in the litter layer, when compared to a site not affected by air pollution.

There can be little doubt that, as a result of the stress caused by pollution in urban areas, the carrying capacity of the soil is greatly reduced. Also this

Table 8.4 Range of microbial characteristics from a metal-polluted soil (Ohya *et al.* 1988)

	Bacteria ($\times 10^{-6}\,g^{-1}$ soil)	Fungal spores ($\times 10^{-4}\,g^{-1}$ soil)	Biomass C ($mg\,g^{-1}$ soil)	CO_2 evolution (mg $100\,g^{-1}$ soil) With substrates	Without substrates
Range	0.33–54	1.6–37	33–439	80.5–1 080	1.5–141
Mean	20	9.3	168	569	42.5

may lead to unusual niches being created, with a community structure unique to urban areas.

TREATMENTS FOR ENCOURAGING SOIL BIOTA

There are various examples of treatments being investigated to encourage plant growth and as a byproduct an increase in the soil biomass. These examples come mainly from attempts by the mineral extraction industries to revegetate reinstated areas.

Cundell (1977) was the first author to stress the importance of microorganisms in the reestablishment of a functioning ecosystem in disturbed soil. He reviewed the limited work done in this area and drew heavily on material from undisturbed ecosystems. This study of western Colorado involved spent shale wastes and the overburden from lignite strip-mine areas in North Dakota as substrates. He reiterated that soil organic matter formation, nitrogen fixation and transformation, and modification of adverse edaphic factors were all microbially-mediated processes, vital for soil formation, and indicated ways to encourage them. Fertilization, mulching, seeding, inoculation of the rhizosphere of perennial grasses with free-living, nitrogen-fixing heterotrophs were all reviewed, but the major conclusion was that a more detailed knowledge of the microbial processes involved in soil formation and plant growth was required. This review seems to have stimulated other work.

Fresquez & Lindemann (1982) carried out a comparison between strip-mined areas restored with topsoil and spoil. The objective was to see whether the latter developed a functional microbial community. A variety of amendments was made to the spoil, separately and in combination, including inoculation with topsoil, addition of alfalfa hay and fertilizer, and addition of gamma-irradiated (and therefore sterile) sewage sludge. All of these were carried out as greenhouse pot experiments and were planted with *Bouteloua gracilis* and then replanted to *Atriplex canescens*. They found that topsoil inoculation alone did not increase numbers of bacteria, fungi, ammonia oxidizers (nitrifiers) or free-living nitrogen fixers. The dehydrogenase activity remained low, and the fungal species diversity narrow. However, both of the organic amendments led to an increase in all measurements of the microbial community. This indicated that a readily available carbon source was more important in stimulating an active and diverse soil microflora than supplying a microbial inoculum. Furthermore, the authors reported an investigation of four areas in the same study: a reclaimed area, a spoil bank, a topsoil stockpile and an undisturbed control, allowing comparison of both type of substrate and disturbance. They found that the undisturbed and reclaimed areas had similar microbial numbers, and fungal genera diversity. The reclaimed area had higher nitrifier numbers and lower dehydrogenase activity than the other areas. The topsoil stockpile and spoil bank had generally lower microbial numbers and dehydrogenase activity than the control, but the topsoil stockpile had higher numbers of nitrifiers. They concluded by recommending that both spoils and stockpiled topsoils should be mixed with fresh topsoil and organic amendments to encourage soil microbial development.

In 1984 Stevenson and co-workers described the results of an investigation into amending soils in central New York, USA, with sewage sludge and the earthworm *Eisenia foetida*. When the earthworms were allowed to disperse there were limited effects on redox potential, moisture content and numbers of invertebrates. However, when confined to lysimeters this produced intense burrowing and egestion activity resulting in stimulation of microbial respiration, with concomitant increases in the flux of oxygen and carbon dioxide. A similar effect was noted when sludge was applied. This offers a possible way of increasing the biological activity of soils in urban areas, although the economic feasibility and environmental acceptability would need to be investigated further.

Visser (1985), investigated the use of amendments for mine spoils, and found that the numbers of bacteria and actinomycetes, total fungal mycelium and microbial biomass carbon increased in the order: NPK fertilizer < sewage sludge < peat. The peat also had the greatest effect in increasing total organic matter and nitrogen. Planting with either slender wheatgrass or white spruce also increased microbial biomass carbon after 27 months on the subalpine site and after 39 months on the oil sand tailings. There were also effects on the rate of decomposition of a variety of litters, with Alsike clover leaves decomposing first followed in order by, grass leaves, clover stems and grass stems. This could be an advantage in the early stages of microbial recovery on mine spoils, as a constant supply of nutrients for primary production could be made available by the rapidly decomposing fraction, whereas the slowly decaying fraction would perhaps contribute to the accumulation of stable organic matter. Growing grass and incorporating its litter into the soil is very important as part of a management regime, either mechanically or preferably by grazing organisms. Visser concluded by indicating the need for this type of organic amendment on restored sites, and for cognisance of differing rates of litter decomposition by those responsible for designing the restoration strategy.

The role of earthworms in the rehabilitation of disturbed sites was investigated by Scullion & Ramshaw (1987) who examined the effects of a number of different manurial treatments on earthworm activity in a restored grassland and an undisturbed control. They found that casting activity could be significantly increased by the use of top-dressed poultry manure in both the worked and unworked land. This concurs with Blenkinsop's (1957) observation that keeping poultry on worked areas led to their successful restoration.

BIOLOGICAL PROPERTIES OF URBAN SOILS/
MATERIALS

Zimny & Zukowska-Wieszczek (1984a) investigated the effect of vegetation type on soil microbial activity as measured by the dehydrogenase assay, in soils of Warsaw, Poland. They recorded the lowest activity in soils taken from around trees growing in pavements along busy streets, and this activity was not affected by a grass cover. The highest activities were recorded in soils in allotment gardens, especially under roses and tomatoes, whereas soils in urban parks fell in the middle of the range of activities recorded. This is perhaps not surprising as there is a strong association with high organic matter and dehydrogenase activity in soils, and organic amendments tend to form a large part of many allotment-holders management strategies, although the authors offered no discussion of their results.

In a study of urban lawns in Warsaw, Poland, Zimny & Zukowska-Wieszczek (1984b) reported that park lawns sheltered by trees and bushes had higher enzyme activities than streetside lawns without any tree or shrub cover. It may be that the low activities in the streetside lawns could be as a result of pollution from car exhausts and lack of organic matter input from trees, leading to a reduced carrying capacity.

Bird & Brisbane (1988) found that bacteria in vineyard soils adjacent to the towns of Loxton and Cooltong in South Australia, significantly reduced the reproductive potential of root-knot nematodes added to fresh soils. In all cases where this inhibition occurred *Pasteuria penetrans* was found in adult females without egg masses, and there appeared to be no inhibitory effects of other bacteria isolated from these soils. It was suggested that *P. penetrans* may be used as a biological control agent for these nematodes. It is important to note that the source of soil for use in urban areas can have an impact on community structure.

Calvo *et al.* (1984) investigated the occurrence of keratinophilic fungi in soils from the city of Barcelona, Spain. Fifty samples were taken and hair-baiting and dilution methods were used to isolate the fungi. The number of spores recorded ranged from 10×10^4 to 10×10^7 per gram of soil; this range is similar to that found in agricultural and 'natural' soils. The genera most commonly isolated by the dilution method were *Penicillium*, *Mucor*, *Aspergillus* and *Cladosporium*, whereas *Cephalosporium* was the genera most commonly isolated by the hair-baiting method. The high occurrence of these species is consistent with the soils being disturbed in a similar way to strip-mined areas (Visser *et al.*, 1983). This lends weight to the use of research carried out on strip-mined areas as useful sources of information for application to some soils from urban areas. The authors also noted that there was a high occurrence (8%) of *Microsporium gypseum* in the samples; this is clinically important because it is implicated in the occurrence of body ringworms.

Alvarez *et al.* (1986) also investigated the occurrence of dermatophytes and geophilic-keratinolytic fungi in the parks, squares and allotments of an urban area, Rosario City, Argentina. They isolated *Trichophyton ajelloi* and *Microsporium gypseum* and found highest numbers associated with argillaceous and sandy argillaceous soils, but no explanation of this was offered.

Bajwa & Jefries (1986) surveyed the keratinophilic fungi population from the Detroit metropolitan area. Of the 100 soil samples taken 92% yielded a total of 172 isolates. Here, *Trichophyton ajelloi* was the most frequently occurring, being found in 58% of the soils. *Microsporium gypseum* occurred in 33% of the soils, almost four times greater than the frequency recorded by Calvo *et al.* (1984), which suggests that the public health risk was higher in Detroit than Barcelona.

In contrast Filipello-Marchisio (1986) examined an unusual habitat, but one common to many urban parks, namely children's sand-pits. The author was particularly interested in the occurrence of keratinophilic fungi as these could be potential pathogens. Twenty-eight sandpits in Turin, Italy, were sampled and 57 fungal species isolated, of which 52% showed keratinolytic activity. The most commonly isolated species are shown in Table 8.5. These all show keratinolytic activity, but perhaps more interesting ecologically the species composition is quite different to that found by Calvo *et al.* (1984). This suggests that the organic enrichment peculiar to sand-pits, such as skin, scabs, feathers, scales and faeces, has led to a distinct ecological habitat. It must also be noted that the abundance of *Microsporium gypseum* was similar to that found by Calvo *et al.* (1984).

Braun & Beck (1986) carried out a survey of pseudoscorpions in a municipal forest at Ettlingen, West Germany; 95% of the total of 3777 caught were of the one species *Neobisium carcinoides*, and the remaining 5% of the species *Chthonius tetrachelatus*. The mean monthly population density was estimated as 45 individuals per square metre and the fresh weight biomass as 27.5 mg m^{-2}. During the period of study (1977–1984) the fluctuation in these was up to a factor of six-fold, although this is very small in comparison to values obtained in temperate grassland soils (cf. Table 8.2). They found that there

was a marked decline in population activity in response to chemical pollutants, such as pentachlorophenol, which the authors ascribed to an already low population density.

Orchard *et al.* (1987) reported on the efficacy of an 'organic liquid amendment' with regard to improving a sports turf located in a local authority ground in the city of Dunedin, New Zealand. Although the amendment produced no significant effect this work allows comparison with soils from undisturbed areas.

Weigmann & Kratz (1987) measured the numbers of oribatid mites in urban and forest areas of West Berlin, West Germany. They found fewer mites in the urban as compared to the forest areas, ascribing this to increased air pollution, isolation of habitats, and increased relative aridity in the urban areas. They suggested that oribatid mites may be useful indicators of soil pollution.

Public health risks

Aside from the literature on the potentially pathogenic keratinolytic and keratinophilic fungi outlined previously, there is some evidence available as to the health risk from soils in heavily used public areas such as parks.

Theis *et al.* (1978) found helminth ova in 1.6% of soil samples and 25.8% of sludge samples from 12 urban areas in the USA. The most prevalent ova were from *Ascaris lumbricoides* followed closely by *Toxocara* sp. The suggestion was made that this was due to pet animal faecal material finding its way into municipal sewer systems.

Giammanco *et al.* (1984) examined soil from the public gardens of 56 towns in the Catania district of Sicily, finding salmonella bacteria in 3.7% of them.

Costa & Camillo-Coura (1985) investigated the link between geohelmintiasis and soil physico-chemical characteristics in urban areas in the State of Pariaba, Brazil. They found that nematode larvae developed best in sandy soils with low P, Ca and Mg, moderately acid to neutral pH, moderate to small amounts of organic matter, an isotherm of 21°C, precipitation between 1200 and 1600 mm annually, and which were well vegetated. In areas with little or no vegetation and with an isotherm of 27°C and less than 800 mm of precipitation annually, larvae developed less frequently.

Table 8.5 Frequency of fungal species in children's sand-pits in Turin, Italy (from Filipello-Marchisio, 1986)

Species	Frequency of occurrence in sand-pits (%)
Aphanoascus terreus	78.5
Aphanoascus fulvescens	67.8
Trichophyton ajelloi	60.7
Chrysosporium tropicum	50.0

Goulart *et al.* (1986) reported on pathogenic fungi isolated from leisure areas in Rio de Janeiro, Brazil. The authors commented on the paucity of information in this field, and described a study of 500 soil samples from a total of 100 districts within the municipality. Approximately 30% of the 352 isolates were identified as being dermatophytes parasitic on man, and warned of the hitherto unrecognized public health dangers of leisure areas. Of particular concern was the fact that 39 pathogenic species were isolated from pleasure beaches.

Discussion and conclusions

The literature concerning the biology of urban soils *in situ* is somewhat limited. Most reports appear to indicate that there are reduced numbers of organisms, coupled with reduced biomass and species diversity. This is greatly influenced by the type of soil being investigated, but generally those soils which have had the greatest disturbance or pollution have the poorest inventory of organisms. This is not, perhaps, surprising as under conditions of stress the carrying capacity of a soil will be reduced, and conditions of disturbance will tend to favour opportunistic organisms. Where both conditions exist simultaneously it is unlikely that any permanent residents of the soil will be found. This may not necessarily be a problem in soils which are fertilized annually as part of their maintenance regime, but in areas of low maintenance and heavy use there could be a risk of structural deterioration, loss of vegetation and subsequent erosion.

Several issues need to be addressed for future research work. A coherent classification of soils in urban areas is required (see Chapter 2). For the purposes of investigating soil biology a simple hierarchical classification would suffice, based on the origin of the soil, and its history in terms of contamination or disturbance.

Within these classifications, numbers of organisms, biomass relationships and species diversity need to be elucidated, in relation to locational and soil use information and soil physico-chemical characteristics. For example, a basic survey of parklands in urban centres could be a starting point, as the numbers and species diversity of the soil community could be related to the level of recreational pressure and contamination. Particular attention needs to be paid to the role of the microbial community as an integrated indicator of soil conditions. This type of survey could then be extended to roadside verges, gardens, allotments and derelict land of various classes, and allow comparison with agricultural and natural areas. Particular problems such as compaction, disturbance and pollution could be investigated in relation to community size and structure.

This would not only create a basic biological database of urban areas but also shed some light on the responses of soil organisms to stress and disturbance. This field of research is currently expanding and studies of urban areas would make a useful addition, as some niches exist which are unique to urban areas.

Once this basic work has begun it will be possible to devise and test treatment strategies for establishing functional soil communities in newly imported materials, or to re-establish such communities in degraded soils, when such as end-point is desired. Recommendations may then be made as to the type of material suitable for use in such areas. There would then also be scope for using soil organisms as indicators of the degree of degradation in any particular area. Up to now the criteria employed have been solely physico-chemical, although this situation is currently under review.

References

Alvarez, D.P., Luque, A.G. & De-Bracalenti, B.C. (1986). Influence of ecological factors on isolation of dermatophytes and geophilic-keratinolytic fungi. *Revista Latino Americana De Microbiologia* **28**, 351–354.

Armstrong, M.J. & Bragg, N.C. (1984). Soil physical parameters and earthworm populations associated with opencast coal working and land restoration. *Agriculture, Ecosystems and Environment* **11**, 131–143.

Bajwa, P.S. & Jefries, C.D. (1986). The keratinophilic fungi in soils from the Detroit metropolitan area, Michigan, USA. *Mykosen* **29**, 267–271.

Bird, A.F. & Brisbane, P.G. (1988). The influence of *Pasteuria penetrans* in field soils on the reproduction of root-knot nematodes. *Review of Nematology* **11**, 75–82.

Blenkinsop, A. (1957). Some aspects of the problem of the restoration of opencast coal sites. *Planning Outlook* **4**, 28–32.

Braun, M. & Beck, L. (1986). On the biology of a beech wood soil, 9. The pseudoscorpions. *Carolinea* **44**, 139–148.

Calvo, A., Vidal, M. & Guarro, J. (1984). Keratinophilic

fungi from urban soils of Barcelona, Spain. *Mycopathologia* **85**, 147.

Costa, W.D. & Camillo-Coura, L. (1985) Edaphological conditions of the distribution of geohelmintiasis in the municipalities of Alhandra serraria and Aguiar in the State of Paraiba, Brazil. *Ciencia Cult Saude* **7**, 7–13.

Craul, P.J. (1985). A description of urban soils and their desired characteristics. *Journal of Arboriculture* **11**, 330–339.

Cundell, A.M. (1977). The role of micro-organisms in the revegetation of strip-mined land in the western United States. *Journal of Range Management* **30**, 299–305.

Daft, M.J. & Nicolson, T.H. (1974). Arbuscular mycorrhizae in plants colonizing coal wastes in Scotland. *New Phytologist* **73**, 1129–1138.

Filipello-Marchisio, V. (1986). Keratinolytic and keratinophilic fungi of childrens sandpits in the city of Turin. *Mycopathologia* **94**, 163 172.

Fresquez, P.R. & Lindemann, W.C. (1982) Soil and rhizosphere micro-organisms in amended coal mine spoils. *Soil Science Society of America Journal* **46**, 751–755.

Fresquez, P.R., Aldon, E.F. & Lindemann, W.C. (1987). Enzyme activities in reclaimed coal mine spoils and soils. *Landscape and Urban Planning* **14**, 359–367.

Fritze, H. (1988). Influence of urban air pollution on needle litter decomposition and nutrient release; a comparison of *Pinus sylvestris* and *Picea abies* L. Karst. *Scandinavian Journal of Forest Research* **3**, 291–298.

Giammanco, G., Marranzano, M. & Giannino, L.R. (1984). The soil of public gardens as salmonella reservoir. *Igiene Moderna* **82**, 762–765.

Golovacheva, R.S., Rozanova, E.P. & Karavaiko, G.I. (1986). Thermophilic bacteria of the sulphur cycle from the corrosion zones of steel constructions in the municipal heating system and soil. *Mikrobiologia* **55**, 105–112.

Goulart, E.G., Lima, S.M.D.F., Carvalho, M.A., Oliveira, J.A.D., Jesus, M.M.D., Campos, R.E. & Cosendey, M.A.E. (1986). Pathogenic fungi isolated from the municipality of Rio de Janeiro, Brazil. *Folia Medica* **93**, 15–20.

Grime, J.P. (1979). *Plant Strategies and Vegetation Processes*. John Wiley & Sons, Chichester.

Hedrick, H.G. & Wilson, H.A. (1956). The rate of carbon dioxide production in a strip-mine spoil. *Proceedings of the West Virginia Academy of Science* **28**, 11–15.

Hersman, L.A. & Klein, D.A. (1979). Retorted oil shale effects on soil microbiological characteristics. *Journal of Environmental Quality* **8**, 520–524.

Hunter, B.A., Johnson, M.S. & Thompson, D.J. (1987). Ecotoxicology of copper and cadmium in a contaminated grassland ecosystem. II. Invertebrates. *Journal of Applied Ecology* **24**, 587–600.

Marrs, R.H., Granlund, I.H. & Bradshaw, A.D. (1980).

Ecosystem development on reclaimed china clay wastes. IV. Recycling of above-ground plant nutrients. *Journal of Applied Ecology* **17**, 803–813.

Miller, R.M. & Cameron, R.E. (1978). Microbial ecology studies at two coal mine refuse sites in Illinois. *Argonne National Laboratory Report* **ANL/LRP-3**.

Miller, R.M. & May, S.W. (1981). Staunton 1 Reclamation Demonstration Project, Soil microbial structure and function: Final Report. *Argonne National Laboratory Report* **ANL/LRP-11**.

Neuhauser, E.F., Loehr, R.C., Milligan, D.L. & Malecki, M.R. (1985). Toxicity of metals to the earthworm *Eisenia foetida*. *Biology and Fertility of Soils* **1**, 149–152.

Ohya, H., Fujiwara, S., Komai, Y. & Yamaguchi, M. (1988). Microbial biomass and activity in urban soils contaminated with Zn and Pb. *Biology and Fertility of Soils* **6**, 9–13.

Orchard, V.A., Ross, D.J., Ross, C.W., Rankin, P.C., Reynolds, J. & Hewitt, A.E. (1987). Some microbiological, chemical and biological properties of two sports ground soils after treatment with an enzyme conditioners. *New Zealand Journal of Experimental Agriculture* **15**, 163–176.

Read, D.J. & Birch, C.P.D. (1988). The effects and implications of disturbance of mycorrhizal systems. *Proceedings of the Royal Society of Edinburgh* **94B**, 13–24.

Richards, B.N. (1987). *The Microbiology of Terrestrial Ecosystems*. Longman Scientific and Technical, New York.

Rushton, S.P. (1986a). Earthworm populations on pasture land reclaimed from open-cast mining. *Pedobiologia* **29**, 27–32.

Rushton, S.P. (1986b). The effects of soil compaction on *Lumbricus terrestris* and its possible implications for populations on land reclaimed from open-cast coal mining. *Pedobiologia* **29**, 85–90.

Schafer, W.M., Nielson, G.A. & Nettleton, W.D. (1980). Minesoil genesis and morphology in a spoil chronosequence in Montana. *Soil Science Society of America Journal* **44**, 802–807.

Schramm, J.R. (1966). Plant colonization on black wastes from anthracite mining in Pennsylvania. *Transactions of the American Philosophical Society* **56**, 194.

Scullion, J. & Ramshaw, G.A. (1987). Effects of various manurial treatments on earthworm activity in grassland. *Biological Agriculture and Horticulture* **4**, 271–281.

Scullion, J., Mohammed, A.R.A. & Ramshaw, G.A. (1988). Changes in earthworm populations following cultivation of undisturbed and former opencast coal-mining land. *Agriculture, Ecosystems and Environment* **20**, 289–302.

Smeltzer, D.L.K., Bergdahl, D.R. & Donnelly, J.R. (1986). Forest ecosystem responses to artificially induced soil compaction. II. Selected soil micro-organism populations. *Canadian Journal of Forest Research* **16**, 870–872.

Standen, V., Stead, G.B. & Dunning, A. (1982). Lumbricid populations in opencast reclamation sites and colliery spoil heaps in county Durham, UK. *Pedobiologia* **24**, 57–64.

Stevenson, B.G., Parkinson, C.M. & Mitchell, M.J. (1984). Effect of sewage sludge on decomposition processes in soils. *Pedobiologia* **26**, 95–105.

Stroo, H.F. & Jencks, E.M. (1982). Enzyme activity and respiration in mine spoils. *Soil Science Society of America Journal* **46**, 548–553.

Swift, M.J., Heal, O.W. & Anderson, J.M. (1979). *Decomposition in Terrestrial Ecosystems*, Studies in Ecology, Vol. 5, Blackwell Scientific Publications, Oxford.

Theis, J.H., Bolton, V. & Storm, D.R. (1978). Helminth ova in soil and sludge from twelve U.S. urban areas. *Water Pollution Control Federation Journal* **50**, 2485–2493.

Vimmerstedt, J.P. & Finney, J.H. (1973). Impact of earthworm introduction on litter burial and nutrient distribution in Ohio strip mine spoil banks. *Soil Science Society of America Proceedings* **37**, 388–391.

Visser, S. (1985). Management of microbial processes in surface mined land reclamation in western Canada. In Tate R.K., III & Klein, D.A. (eds), *Soil Rehabilitation, Processes, Microbial Analyses and Application*, pp. 203–241. Dekker, New York.

Visser, S., Griffiths, C.L. & Parkinson,D. (1983). Effects of surface mining on the microbiology of a prairie site in Alberta, Canada. *Canadian Journal of Soil Science* **63**, 177–189.

Weigmann, G. & Kratz, W. (1987). Oribatid mites in urban zones of West Berlin, West Germany. *Biology and Fertility of Soils* **3**, 81–84.

White, C.S. & McDonnell, M.J. (1988). Nitrogen cycling processes and soil characteristics in an urban versus rural forest. *Biogeochemistry* **5**, 243–262.

Wilson, H.A. (1965). The microbiology of strip mined spoil. *West Virginia Agricultural Station Bulletin* **506T**, 44pp.

Wilson, H.A. & Stewart, G. (1955). Ammonification and nitrification in a strip-mine spoil.*West Virginia University Agricultural Experiment station Bulletin* **379T**.

Vilson, H.A. & Stewart, G. (1956). The number of bacteria, fungi and actinomycetes in some strip mine spoils. *West Virginia University Agricultural Experiment Station Bulletin* **388T**.

Zimny, H. & Zukowska-Wieszczek, D. (1984a). Dehydrogenase activity depending on the kind of urban greens. *Polish Ecological Studies* **9**, 113–122.

Zimny, H. & Zukowska-Wieszczek, D. (1984b). Enzymatic activity in soils of urban lawns depending on sources of degradation. *Polish Ecological Studies* **9**, 123–130.

9 Soils and vegetation in urban areas

H.J. ASH

Introduction

Urban botany is a relatively new branch of botanical science: until recently botanists sought the country-side, horticulturalists kept to their gardens, and the rest of the urban scene was left largely to the local authority. As a result there are considerable gaps in our knowledge and understanding of urban vegetation. Attempts have been made to classify urban vegetation, both on a simple habitat basis, extending the NCC Phase I classification (NCC, no date; St Helens WAG, 1986), and on a more detailed basis by Shimwell in his *Conspectus of Urban Vegetation* (1983) – see Appendix. The National Vegetation Classification (NVC, no date) has yet to cover many urban habitats, but has contributed, along with the above references, to the following section.

Urban vegetation types

THE PIONEERS

The influence of urban soils on their vegetation is often strongest in the early successional stages. The key soil factors appear to be nutrient availability, especially nitrogen and phosphorus, and water supply (Table 9.1). Also important is the pool of species available to each site, both as stored propagules on site (seed, roots, tubers) and those arriving from outside.

Thus the community found where both nutrients and water are plentiful, as in eutrophic muds and silts around sewage farms and some reservoirs, is marked by an abundance of the Chenopodiaceae. It has few species in common with those found where nutrients are plentiful but water supply no more than adequate, e.g. in gardens and on active household waste tips (Table 9.1). However, both these communities are regularly disturbed, for example by fluctuating water levels or cultivation, which favours annual species.

Where disturbance is less, but supplies of water and nutrients also limited, perennials and annuals are found as early colonizers. This group includes a wide variety of substrates. Demolition sites, where soil is mixed with brick rubble, often support large populations of the yellow crucifers, while hairy bitter-cress (*Cardamine hirsuta*) favours calcareous or cinder soils, spurrey (*Spergularia arvensis*) sandy soils, and scarlet pimpernel (*Anagallis arvensis*) with speedwells (*Veronica* spp.) show a preference for acidic railway tracks and cinders. Undisturbed sandy soils sometimes develop deep carpets of the firmoss (*Polytrichum juniperum*), often with lichens such as *Cladonia pyxidata*, *C. chlorophaea* and *Peltigera* spp. Occasionally two recent introductions, gallant-soldier (*Galinsoga ciliata*) and Canadian fleabane (*Conyza canadensis*) dominate light sandy soils, but they are not yet found throughout Britain.

Severe deficiency in nutrients or water, or both, usually favours lower plants, especially where there is no depth of loose material to act as a soil. Mosses, liverworts and lichens can obtain sufficient nutrients from rainfall and runoff, and some can withstand desiccation and high temperatures. Industrial wastes, which have severe edaphic problems often including extreme pH, are dealt with later. Among the lower plant communities, the pH of the substrate has a marked effect on the species present (Table 9.1). Lower plants, in particular the lichens, are sensitive to air pollution, and with cleaner air are increasing in many urban areas, especially on calcareous substrates.

Walls may harbour scattered higher plants, such as wallflower (*Cheiranthus cheiri*), an old introduction which favours limestone walls, and various native ferns (*Asplenium* spp., *Dryopteris felix-mas*), plus a variety of weedy species. More recent walls (less than 100 years old) use harder mortar which is more resistant to decomposition and will probably prove more difficult to colonize (Woodell, 1979). Similar communities sometimes develop on dry, disturbed soils, e.g. under elevated motorways.

In all these communities the species present, and their abundance, is heavily dependent on the supply of propagules. There is often considerable 'founder

Nutrients	Water	Habitat	Species
Abundant	Abundant	Sewage works	*Atriplex prostrata* (orache) *Bidens* spp. (bur-marigolds) *Chenopodium rubrum* (goosefoot) *Polygonum* spp. (knotgrasses)
Abundant	Adequate	Gardens	*Bromus sterilis* (barren brome) *Chenopodium album* (fat-hen) *Hordeum murinum* (wall barley) *Senecio vulgaris* (groundsel) *Stellaria media* (chickweed) *Veronica* spp. (speedwells)
Limited	Limited	Brick rubble	Perennials: *Agrostis capillaris* (bent-grass) *Rumex crispus, R. obtusifolius* (docks) *Tussilago farfara* (cocksfoot) Annuals: *Diplotaxis* spp. (rockets) *Senecio squalidus* (Oxford ragwort) *Sinapis arvensis* (charlock) *Sisymbrium* spp. (hedge mustard) *Tripleurospermum inodorum* (mayweed)
Deficient	Adequate	Viaducts	Thalloid liverworts especially *Marchantia polymorpha*, scattered mosses
Deficient	Deficient	Tarmac (neutral/acid)	Mosses: *Bryum argenteum* *Ceratodon purpureus*
		Brick (neutral/ alkaline)	Mosses: *Funaria hygrometrica* *Leptobryum pyriforme* Liverwort: *Lunularia cruciata*
		Concrete (alkaline & calcareous)	Mosses: *Barbula* spp. *Bryum* spp. *Tortula muralis* Lichens: *Xanthoria* spp. *Candelariella* spp.

Table 9.1 Pioneer plant species of urban soils

effect' – the first colonizers, if they can reproduce at all, have the opportunity to establish themselves unopposed. In consequence, there is great variation both between sites with apparently similar edaphic conditions, and also in time as the propagule supply changes, existing colonists increase their populations and competition between individuals begins.

THE SUCCESSORS

Pioneer communities may last for a few years, or, where environmental stress is greater, persist for decades. Where water supply is abundant, the succession proceeds to marshland, but in drier habitats, as found in most urban situations, grasses and herbs are the norm. The communities which develop fall into two broad groups: those of moderately to very fertile soils, and those of infertile, usually nitrogen-deficient, substrates such as brick rubble. (Substrates with extreme edaphic conditions are dealt with in the next section.)

The first group consists of rough grasslands, usually species-poor. In the metropolitan boroughs of St Helens and Knowsley, this habitat accounts for a quarter of the urban greenspace (10% of the total urban area). It develops rapidly (within five years) on most neutral, mesotrophic soils of moderate to good fertility when mowing or grazing is stopped, and is found on abandoned farmland, older refuse tips (Gemmell, 1988), amenity grassland if left unmown, motorway banks and many plots of untended land. It is classified by the NVC under its dominant grass, false-oat (*Arrhenatherum elatius*). Cocksfoot (*Dactylis glomerata*) is often co-dominant. Both form large tussocks and masses of dead foliage which smother many smaller species. The big umbellifers, cow-parsley (*Anthriscus sylvestris*) and hogweed (*Heracleum sphondylium*), are the most obvious associates, together with Yorkshire fog-grass (*Holcus lanatus*), docks (*Rumex*), thistles (*Cirsium*) and couch-grass (*Elymus repens*), and sometimes other herbs such as knapweed (*Centaurea nigra*) and ox-eye daisy (*Leucanthemum vulgare*). Two garden escapees, Michaelmas daisy (*Aster novi-belgii*) and goldenrod (*Solidago canadensis*) have established themselves in many urban examples.

Four of the subcommunities recognized by the NVC occur in urban areas. One (*Festuca rubra*) is mentioned in the section on 'man made' habitats; the others are dependent on variations in soil pH, nutrients (N and P) and drainage. The *Centaurea nigra* subcommunity is associated with rich meso-trophic soils, including calcareous ones: some of the best examples are in churchyards. It is the most species-rich type, characterized by knapweed (*Centaurea nigra*), ox-eye daisy (*Leucanthemum vulgare*) and frequently ragwort (*Senecio jacobaea*), with

many smaller grasses and herbs. The other three subcommunities (*F. rubra*, *Filipendula ulmaria* and *Urtica dioica*) are associated with brown earths of neutral pH, which in urban habitats are frequently stony, shallow or compacted and clayey. The meadowsweet (*Filipendula ulmaria*) subcommunity develops where moisture collects, especially in roadside ditches, and is uncommon in the urban area. The stinging nettle (*Urtica dioica*) subcommunity, which usually has large amounts of the big umbellifers and cleavers (*Galium aparine*) as well as nettles, is associated with areas of high nutrient status. On some sites the nettles almost form a monoculture.

The NVC states that unmown *Arrhenatheretum* is 'rapidly invaded by shrubs' and forms a temporary stage in the succession to woodland. On many urban sites this does not appear to be happening. On a former railway triangle in Merseyside extensive areas of *Arrhenatheretum* on former farmland show little shrub colonization after 100 years, despite extensive tree and shrub colonization of the lines and sidings. A possible reason may be the frequency of fires, feeding on the substantial amounts of dead material always present in this community. Such fires have little effect on the community itself, especially when they occur outside or early in the growing season (NVC), but summer fires encourage rosebay willowherb (*Chamerion angustifolium*), helping to maintain its prominence in urban areas. Fire also kills tree seedlings, holding the succession at a grassland stage. However, a more general reason may be that the dense sward is very difficult for small-seeded species to invade, and the common pioneer species of urban areas are those with small, light seeds, e.g. the birch and willow that have colonized the former tracks. A few young hawthorn and oak do occur on this site, and nearby, abandoned farmland with hedges is rapidly becoming hawthorn scrub. These suggest that larger-seeded species can invade the *Arrhenatheretum* and that once again the reason for their absence in the urban area lies in seed availability rather than in the soil or vegetation type.

The *Arrhenatheretum* is not a valued community in towns. It is regarded as 'untidy', unattractive, a sign of neglect, and is prone to abuses such as fly-tipping, fires, unauthorized camping, etc. Its only profitable uses appear to be limited amounts of dog-walking, and for children's play. The latter is important (there are very few informal areas close to home for today's urban children to make their own worlds), but a more varied community with trees and shrubs would offer many more possibilities.

The coarse grasslands also have little wildlife value, plant or animal. In Grime's 'humpback' model of species diversity versus biomass (Fig. 9.1), the very large accumulations of biomass in rough grassland put it well above the 'hump' of high species diversity. If the model holds good, then some means of reducing the biomass is needed to create a more diverse grassland, whether by changing the species composition, mowing/grazing or trying to reduce soil fertility. Cases where grazing has been resumed on neglected pastures (NVC) suggest that the *Arrhenatheretum* can be converted to a finer-leaved sward with no tussocks within five years of sheep grazing. Whether the same could be achieved by mowing, using suitable equipment which will pick up the cuttings, is not known. Techniques for introducing species to increase the attractiveness of the community are still in the experimental stage. One major problem is that much rough grassland exists because there is no money available to maintain it – and consequently none to try to alter it.

The succession on less-fertile materials may be typified by that on brick rubble. Demolition sites are

currently a common urban feature, therefore brick rubble, mixed with varying amounts of soil, is one of the most abundant urban substrates. Its plant succession has been catalogued in Sheffield by Gilbert (1983). The soil is deficit in N, but usually (depending on the clay used for the bricks) has adequate P, is stony and often compacted (Table 9.2). pH is usually ≥ 7 (Bradshaw & Chadwick, 1980). The first colonizers in Sheffield are annuals and short-lived perennials: Oxford ragwort (*Senecio squalidus*), knotgrass (*Polygonum aviculare*), hastate orache (*Atriplex prostrata*) and fat hen (*Chenopodium album*). Within three to six years, tall herbs take over. Colonization by clovers (usually *Trifolium repens*) tends to speed up this stage.

The tall herbs include rosebay willow-herb (*Chamerion angustifolium*) tansy (*Tanacetum vulgare*), wormwood (*Artemisia absinthum*), mugwort (*A. vulgaris*) and occasionally goldenrod (*Solidago canadensis*) and Jacob's ladder (*Polemonium caeruleum*). Two features of these assemblages are interesting. First, they include many genera which were widespread in Britain 10 000 years ago, as the land was recolonized after the last Ice Age. There is considerable similarity in the conditions; intermittent disturbance, low competition between species, and disturbed soils of relatively high pH, low nitrogen content, often well-drained but frequently compacted (Sukopp *et al.*, 1979). Second, many of the species involved are aliens, though often long-established ones (tansy, goat's rue, wormwood, Oxford ragwort, etc.). Explanations for these are more difficult. Availability of seed will be one factor: some species have wind-blown seeds (*Artemisia* spp., rosebay willow-herb, ragwort), others are currently grown in gardens (goldenrod) or may have lain dormant in the soil from previous fashions in cultivation (tansy, goat's rue). There could be other reasons, e.g. to do with climate. Sukopp *et al.* (1979) and Henke & Sukopp (1986) reported similar patterns in the much drier climate of Berlin, with many ruderals including aliens among the colonizing species, along with significant numbers of meadow and dry-grassland species. Even in Central European cities some of the same species recur, e.g. *Aster*, goldenrod (*Solidago canadensis*) (Kunick, 1982).

Grasses gradually invade the brick rubble succession, so that after about ten years, sites appear as grassland with clumps of tall herbs. Woody species

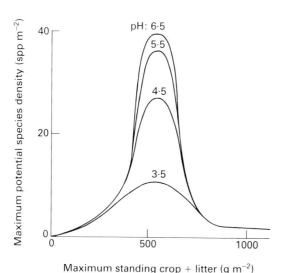

Fig. 9.1 The relation between species richness and the seasonal maximum dry weight of crop and litter in soils with different surface pH for British herbaceous vegetation (from Grime, 1979).

Table 9.2 Characteristics of soils of urban clearance sites (from Dutton & Bradshaw, 1982)

	Bulk Density (t m^{-3})	N Total (ppm)	Mineralizable N (ppm)	Available P (ppm)	K (ppm)	Mg (ppm)
Urban sites	1.7	559	4	26	106	233
Garden soil	1.2	1 027	119	30	83	66
Satisfactory levels	1.5		50	10	80	50

also appear, and gradually the grasslands become scrub and then, some 40 years after demolition, incipient woodland. Similar successions could be catalogued for other urban areas, varying only in some of the species involved. Thus in London, early colonizers include mignonette (*Reseda lutea*), viper's bugloss (*Echium vulgare*), marjoram (*Origanum vulgare*) and bromegrass (*Bromus erectus*) (Cole, 1986).

The influence of legumes on both these successions is interesting. Clovers (*Trifolium* spp.) and some other species readily establish in a range of urban soils so long as there are adequate supplies of phosphate, especially on existing, if disturbed, soils where the necessary *Rhizobium* bacteria may be already present. Their ability to fix atmospheric nitrogen gives them an advantage on sites where that element is in short supply, and their activities can greatly accelerate the succession from pioneers to grassland and on to scrub/woodland (A.D. Bradshaw, pers. comm.) Work by Bradshaw and others on china clay waste suggested that a minimum level of nitrogen (about 700 kg ha^{-1} in that ecosystem) was necessary to support shrub and tree growth (Bradshaw & Chadwick, 1980). White clover (*Trifolium repens*) in pastures fixes at least 100–200 kg N ha^{-1} year^{-1}. Communities on disturbed urban soils are less productive, but still make a significant contribution to the accumulation of nitrogen (Dancer *et al.*, 1977). Ironically, clovers tend to be the victim of their own success. In the absence of management, coarse grassland is likely to develop as soil nutrient supply increases, with which white clover is unable to compete, although red clover (*T. pratense*) and vetches (*Vicia* spp.) may persist. Alternatively, a developing woody canopy will create too much shade for legumes to flourish.

A few other grass and herbaceous species can form distinctive communities in urban areas. Bracken (*Pteridium aquilinum*) is infrequent in urban areas, but may form dense stands on dry, acidic soils, e.g. railway embankments, cinder soils on old factory areas. A number of alien species occur widely, most originating as garden escapes, e.g. Japanese knotweed (*Reynoutria japonica*), lupins (*Lupinus polyphyllus*) and the melilots (*Melitotus* spp.). Horseradish (*Armoracia rusticana*) frequently forms extensive stands around allotments and on road verges, although its fruit never ripens in this country. Other aliens are more localized, and what influence soil may have on the distribution of these species is unknown.

On damp soils, tending towards marshland, various gregarious herbs can form dense stands. The common ones include hemp agrimony (*Eupatorium cannabinum*), great hairy willow-herb (*Epilobium hirsutum*) and yellow loosestrife (the native *Lysimachia vulgaris* or the garden escapes *L. punctata* and *L. ciliata*). Various others are occasionally encountered, such as fleabane (*Pulicaria dysenterica*), comfrey (*Symphytum* spp.), hemlock water-dropwort (*Oenanthe crocata*). The latter is normally found adjacent to canals and streams, but there is little knowledge of soil preferences within this group. 'Founder effect' may be important, as dense stands are difficult for other species to invade.

Among the grasses, tall fescue (*Festuca arundinacea*) may form a species-poor, tussocky sward on damp, brackish or calcareous soils, especially those of industrial origin, producing a community which looks superficially like *Arrhenatheretum*. In more 'natural' situations this is a community of upper salt marshes and tidal rivers (NVC). Tufted hair-grass (*Deschampsia caespitosa*) is frequently found, and becomes dominant on permanently moist areas, wherever mineral soils are oligotrophic and periodically anaerobic, and thus inhospitable to many other neutral grassland species (NVC). Examples occur on old rubbish dumps, road verges, edges of pools, areas stripped of topsoil, etc. The tufted hair-grass is commonly associated with Yorkshire fog-grass (*Holcus lanatus*), but few other species can compete and the hair-grass may form almost pure stands, growing to large tussocks of sharp-edged leaves. In flower it is an attractive community, with masses of tall feathery plumes. It produces large amounts of wind-blown seeds and will quickly colonize exposed areas of suitable soil. A subcommunity where oat-grass (*Arrhenatherum elatius*) and cocksfoot (*Dactylis glomerata*) are co-dominant with *D. caespitosa* is common on slightly drier sites such as verges and churchyards, where tufted hair-grass cannot maintain its competitive advantage.

TOWARDS A CLIMAX

In the absence of controls, most areas of the British Isles would eventually return to woodland, and urban land is no exception. However, few urban sites have yet been neglected long enough to progress this far, and many are held in check by occasional mowing or burning. Gilbert (1983) considers that in Sheffield it takes around 40 years to get incipient woodland on brick rubble. This woodland rarely resembles rural examples. Gilbert (1983) quotes the best examples (in Newcastle-upon-Tyne and Birmingham) as composed of ash (*Fraxinus excelsior*), sycamore (*Acer pseudoplatanus*), laburnum (*Laburnum amygdaloides*), hawthorn (*Crataegus monogyna*), Swedish whitebeam (*Sorbus intermedia*), crab apple (*Malus sylvestris*), guelder rose (*Viburnum opulus*) broom (*Cytisus scoparius*) and goat willow (*Salix caprea*). Elder (*Sambucus nigra*) and bramble (*Rubus fruticosus* agg.) sometimes occur. Willows are often the pioneers, as in late-Glacial times, but many species arrive simultaneously, before the sward closes and renders colonization difficult. The species composition of the scrub appears to change owing to the different growth rates of the constituents. All the species involved thrive in the urban environment, tolerating a range of soil conditions, although the calcareous nature of brick rubble is particularly favourable to ash, whitebeam and guelder rose. The accidents of seed availability are again important in determining the species present on any particular site. Some good examples occur in coal-mining areas where old (50–100 years) colliery shale tips

have been colonized by birch/oak woodland (*Betula pendula/Quercus robur*). On very poor soils such as industrial wastes, pioneer trees may be among the early colonizers, the usual species being silver birch (*Betula pendula*) and scrub willows (*Salix caprea, S. cinerea*). Once a closed, grass sward develops these small-seeded species find it very difficult to establish. While nutrient supplies remain poor, the woody species grow very slowly, so that the vegetation functions as a grassland for some decades (Brierley, 1956; Down, 1973; Ash, 1983).

On dry, acidic substrates, gorse (*Ulex europaeus*) and/or broom (*Cytisus scoparius*) thickets may develop (Shimwell, 1983). Ericaceous heath is a rarity in urban areas, confined to some areas of captive countryside (see below) and a few colliery shale heaps. Both communities pose problems in urban situations, because of the fire risk, but are attractive to people and animals.

On damp soils, birch/willow pioneer scrub may be supplemented by taller willows (e.g. *Salix fragilis*), poplars (*Populus tremula*) and sometimes, if seed is available, alder (*Alnus glutinosa*).

In addition to the natural succession, most urban areas contain large numbers of trees along streets, in gardens and over mown grass in parkland ('urban savannah'). These usually bear little relation to soil type, being chosen for their abilities to survive urban life, particularly in smokier days (e.g. London plane *Platanus acerifolia*, locust tree *Robinia pseudoacacia*), or for their attractive features (e.g. horse chestnut *Aesculus hippocastanum* and numerous flowering cherries – *Prunus* cultivars). Many are not native species, as is also the case with the ornamental shrubberies of gardens and parks. One consequence is that the same species tend to occur in most urban areas, a phenomenon also noted for Central European cities by Henke & Sukopp (1986).

Copses and woodlands in urban areas are mostly (except for such recent efforts as Warrington New Town – Tregay & Moffat, 1980; Scott *et al.*, 1986) examples of 'captive' countryside, built around as the town spread. Many are plantations originally made for either timber or game preservation, while some are truly ancient woodlands with soils to match, e.g. Dibbinsdale on Wirral, Merseyside, and Perivale Woods, Ealing, London (see Smyth, 1987).

Captive (or 'encapsulated' or 'relict') countryside has a major role in urban areas, particularly the larger conurbations (Teagle, 1978; Laurie, 1979; Handley, 1983; Cole, 1986). The gazeteer in Smyth (1987) reveals that many urban wildlife areas (or at least, many officially known ones) are relicts of the former landscape, especially in London. Sometimes management has also survived, or is even the reason for their survival as at Yarnton Meads, an Oxford hay meadow. Examples include woods, heaths, ponds, marshes and even meadows, of which London has a surprisingly large number relative to neighbouring counties (LEU, 1988). Some work has been done relating these to soil types (e.g. NVC *Mesotrophic Grasslands* for meadows; Rackham (1980) for woodlands), although past and present management are at least as important in determining species composition. The London Ecology Unit was able to classify most of its meadows according to the National Vegetation Classification, a national system based largely on rural data. They include a few classic, crested dog's tail – knapweed meadows (*Cynosurus cristatus – Centaurea nigra*), and even one flood meadow (*Cynosurus – Caltha palustris*), a very rare type. A new subcommunity had to be inserted in the NVC to cope with the Thames grazing marshes, rich in meadow barley (*Hordeum secalinum*), and some sites failed to fit any NVC category. The most frequent of the latter were areas containing abundant couch-grass (*Elymus repens*), which may have persisted since use of the land for wartime allotments.

Many encapsulated sites are changing as traditional management has lapsed, often being replaced by intense public pressure, plus increased pollution and invasion by alien species from nearby gardens (Peterken, 1986). Woodlands frequently show poor age structure, little regeneration and a declining ground flora (Numata, 1982). They are in urgent need of replenishment – as became abundantly clear in southern England in October 1987, when many over-mature trees fell in a violent storm. Often such sites are in local authority control and there is urgent need for inputs of money and skill, as well as willpower, to manage them properly. Wirral, Merseyside, possesses a number of examples of lowland heath, a nationally rare habitat, yet at least 60% of it is colonized by birch and rapidly turning into woodland, as a result of the cessation of controlled burning (J. Daniels, pers. comm.), presenting the local authority with a major management problem.

Management of encapsulated countryside requires less frequent and intensive care than traditional parks (and is usually cheaper), but it is dependent on skills not always available within a Local Authority, and on work being correctly timed (Scott *et al.*, 1986). It may involve management techniques that are novel in urban areas (Green, 1986). However, such countryside offers considerable psychological benefits and opportunities for community involvement and education (Cole, 1986).

WATER AND WETLANDS

Water is important to the pattern of vegetation in urban areas, (Eaton, 1986). Not all urban waters are so polluted as to be lifeless; indeed, streams entering the urban area may be richer in wildlife than their rural stretches, since they often get less nutrient-rich input, and are not cleared out regularly (St Helens WAG, 1986; Ash, 1988). Unfortunately as the streams penetrate further into the urban area, water quality usually deteriorates, or they vanish underground. Water bodies isolated from the land drainage system, such as canals, ponds, reservoirs and subsidence flashes, are often reasonably clean and support much aquatic and marsh life. Canals may allow the spread of such species along their length (R.P. Gemmell, pers. comm.). Fringed water lily (*Nymphoides peltata*) appears to be spreading along the Leeds–Liverpool canal in Liverpool.

In static and slow-moving waters, the most abundant surface vegetation is the duckweeds (*Lemna minor*, *L. trisulca*), which provide food for grazing waterfowl. Also common are broad-leaved pondweed (*Potamogetan natans*), which prefers waters 1–2 m deep with muddy or silty bottoms rich in nutrients, and amphibious bistort (*Polygonum amphibium*). The latter is tolerant of a range of depths and often flourishes where the water level fluctuates, as at the edges of ponds. Under the surface the Canadian pondweeds (*Elodea* spp.) abound, replacing native species. The most abundant natives are the water starworts (*Callitriche* spp.), which are most noticeable in shallow waters where they form floating rosettes. Both groups play an important role in the ecosystem by oxygenating the water and providing shelter for invertebrates. Less common are the small-leaved pondweeds (*Potamogeton pectinatus*, *P. crispus*), water-milfoils (*Myriophyllum* spp.)

and water lilies (*Nuphar lutea*, *Nymphaea alba*), all of which seem to like clean conditions. In unpolluted, still, calcareous waters the large algae known as stoneworts (*Chara*, *Nitella*) are occasionally found, while in the few clean, fast-flowing rivers and canal locks in urban areas, the water mosses *Fontinalus antipyretica* and *Eurhynchium riparoides* often occur.

In addition to the variations caused by water movement, pH, depth and pollution, some species show regional variations. The hornworts (*Ceratophyllum*) are commonest around London, while water-soldier (*Sratiotes aloides*), which like hornwort sinks in autumn and rises to the surface in spring, is relatively common in Greater Manchester and Merseyside. Water-soldier needs calcareous waters (it sinks by accumulating calcium in the leaves), and is established in a number of flooded marl pits. These pits, dug in the eighteenth century to extract calcareous clay to improve the fields, are very common in large areas of north-west England and some have been 'captured' intact by spreading towns. The species is not native locally, but is often thrown out from garden ponds and readily increases in suitable waters, posing a management problem. In the Midlands and Greater Manchester, the floating water-plantain (*Luronium natans*) is established and apparently spreading, favouring shallow (<1 m deep), slow moving waters with muddy bottoms.

Marshes are one of the fastest-colonizing urban habitats. Untended sites can develop excellent plant and animal life in 10–20 years (Ash, 1983). For example, a site at Kirkby, Merseyside has a plant list of over 100 species, several locally uncommon or rare, after about 30 years. Among the earliest colonizers of marshes, and the most spectacular, may be the marsh orchids (*Dactylorhiza* spp.). A site at Wigan, where pulverized fuel ash (PFA) was tipped into a subsidence flash to just above water level, was supporting a large hybrid swarm within ten years of abandonment.

Large marshes are not a common urban feature, being usually restricted to areas of industrial dereliction or river valleys. However, small marshy patches are extremely common, especially in areas of clay soil or other poor natural drainage. Marsh vegetation may be roughly divided into tall (1–1.5 m) and short (*c.* 0.7 m). The former is more common on the edges of water bodies and in large marshes. It

is dominated by reedmace (*Typha latifolia*), reed canary-grass (*Phalaris arundinacea*) and/or reedgrass (*Glyceria maxima*), sometimes in pure stands, but often as a species-rich mixture with great hairy willow-herb (*Epilobium hirsutum*), stinging nettle (*Urtica dioica*), fiorin (*Agrostis stolonifera*), yellow flag (*Iris pseudacorus*), etc. The soil is normally nutrient-rich, especially where *Typha* dominates. *Phalaris* favours drier sites with fluctuating water tables, while in areas with standing water for most of the year, the above species may be replaced by branched bur-reed (*Sparganium erectum*) and water-plantain (*Alisma plantago-aquatica*). Common reed (*Phragmites australis*) is rarely found in urban areas. It is confined to permanently wet sites with water of high conductivity, and where it does occur is often associated with industrial activity. Tall sedge-marsh, dominated by lesser pond-sedge (*Carex acutiformis*), is also unusual, but where a dense stand does occur, has the distinction of being one of the least penetrable of urban habitats, even to the local children. Where pollution is too great for most native species Himalayan balsam (*Impatiens glandulifera*), introduced as a garden plant in the late nineteenth century, has stepped in, forming a more or less continuous fringe along polluted water-courses.

Low-growing marshes are very common in urban areas, developing wherever surface drainage fails so that soils are permanently moist. They occur on amenity and reclaimed areas, demolition sites and neglected land, as well as on large marshes, on a wide range of soil types. Most examples are species-poor and characterized by rushes. Soft rush (*Juncus effusus*) tends to colonize nutrient-poor, acidic soils, while hard rush (*J. inflexus*) and compact rush (*J. conglomeratus*) favour neutral–alkaline ones, although all three species can frequently be found growing together. Such communities are classified by the NVC *Mesotrophic Grasslands* as *Holcus lanatus – Juncus effusus* rush pasture; Yorkshire fog-grass and fiorin (*Agrostis stolonifera*) are the typical grasses and creeping buttercup (*Ranunculus repens*) the only constant among the few dicotyledons found in this community. The edges of these rush marshes, and other areas with moist but free-draining soils, often have fragmentary stands of fiorin associated either with silverweed (*Potentilla anserina*) and red fescue (*Festuca rubra*) (NVC *Festuca – Agrostis –*

Potentilla inundation grassland), or with marsh foxtail (*Alopercurus geniculatus*) (NVC *A. stolonifera – A. geniculatus*). The latter favours silty, almost neutral soils. Both communities tend to be species-poor, if productive, and vulnerable to poaching if mown or trampled.

Low-growing sedge marshes (*Carex* spp.) are much less common than rush-marshes in urban areas. Their seeds are generally less easily distributed than those of rushes, which have mucilaginous coats that stick to animals, people and machinery. On neutral–acidic soils species-poor sedge-marsh may develop, dominated by common sedge (*Carex nigra*) and occasionally, when there is an available seed source, including cotton-grass (*Eriophorum angustifolium*). More rarely, species-rich examples occur on neutral–alkaline soils, often of marshy, industrial origin. Carnation-grass (*Carex flacca*, *C. panicea*) and jointed rush (*Juncus articulatus*) are the major species, with many herbs including orchids (*Dactylorhiza fuchsii*, *D. praetermissa*, *D. incarnata*, *D. purpurella* and sometimes other genera).

In wetter areas, spreading into the edges of water bodies, may be beds of flote-grass (*Glyceria fluitans*, *G. plicata*), brook lime (*Veronica becca-bunga*), spike-rush (*Eleocharis palustris*), water horsetail (*Equisetum fluviatile*), water mint (*Mentha aquatica*) or water parsnip (*Berula erecta*). All tend to occur as pure stands rather than mixed communities, for which their growth habits (usually rhizomatous) are partly responsible, but little is known of their ecology.

There is also little evidence as to why certain sites develop tall marsh, often attractive and species-rich, while others support only low-growing species-poor communities. It seems probable that the tall marsh occupies the more eutrophic sites. It may also be that tall marsh develops more readily on bare soil (reedmace for instance has a small wind-blown seed), whereas the somewhat larger-seeded rushes can more easily invade existing grassland whose vigour has been reduced by failing drainage. Water levels and water quality may also be involved.

Few urban marshes have existed undisturbed for long enough to show succession beyond the herbaceous stage. Occasional willows (*Salix caprea*, *S. cinerea*, *S. viminalis* or, in Merseyside and Greater Manchester, *S. repens*) suggest that some at least might eventually become willow carr if allowed to do so, and if the willow seeds can find open patches in which to germinate.

'MAN-MADE' HABITATS

Few urban (or rural) habitats are unaffected by man's influence, but his effects are most deliberately applied to amenity vegetation and gardens. In Knowsley Metropolitan Borough, Merseyside, half of the urban greenspace (20% of the total urban area) is amenity grassland, maintained by intensive mowing regimes. Similar figures apply to other areas (St Helens WAG, 1986). This is a uniform and species-poor community, classified by the National Vegetation Classification and Shimwell into only two major types:

1 Ryegrass – meadow grass (*Lolium perenne*, *Poa pratensis*) recreational swards where there is considerable trampling pressure. White clover (*Trifolium repens*) is usually present, often co-dominant, and there is a small range of associated species, e.g. Yorkshire fog-grass (*Holcus lanatus*), cocksfoot (*Dactylis glomerata*), plantains (*Plantago* spp.) and dandelion (*Taraxacum officinale* agg.), with small amounts of fine-leaved grass, bent (*Agrostis capillaris*) and fescue (*Festuca rubra*).

2 Ryegrass – ribwort (*Lolium perenne – Plantago lanceolata*) verges and lawns. This community is commonly created from seed using large amounts of ryegrass and white clover. With age ribwort becomes constant (sometimes abundant especially on slopes), and the same associates appear as in the first type. It is used where trampling is only moderate.

The two types can be difficult to separate in the field, especially when recently mown. Both are commonly sown on bare soil, but may have been created by sowing into existing meadows and pastures. The intensive mowing regimes (6–16 cuts per annum) help to eliminate all but a few associated species. The low ecological interest of amenity grasslands has led to them being dubbed 'green deserts'. They occur over a wide range of almost neutral soils of moderate to high fertility. The key influences in their maintenance are not soil type but seed input and, especially, management. Consequently they impart an appearance of uniformity to urban landscapes countrywide.

Detailed survey of such areas may reveal some variations where patches of different (usually poor) soil or original vegetation have managed to survive. In Kirkby, Merseyside, a few sites retain areas of native bent-grass (*Agrostis capillaris*), quite different in growth form to the agricultural cultivars, and typically associated with sheep's sorrel (*Rumex acetosella*), hawkbit (*Leontodon autumnalis*) and the moss *Dicranella heteromalla*. One site is a reclaimed sandstone quarry, the others examples of farmland 'mown into' amenity grassland with little reseeding. As examples of acidic grassland they are very species-poor, but they do suggest infertile, possibly acidic soils, which would probably be more amenable to species introduction and relaxation of management than normal amenity grassland.

Most unusual was a small area found, during the St Helens Phase I survey, at Newton-le-Willows, where a sandstone quarry had been reclaimed to amenity grassland. An area some 15 m square on the lip had retained an acidic marsh flora including bog pimpernel (*Anagallis tenella*), with a drier edge of heather (*Calluna vulgaris*), despite being mown regularly to 5 cm. Mowing has now been stopped and the development of the site is being monitored with interest.

A relatively new development is the sowing of fine-leaved amenity grasslands, composed of slow-growing cultivars of bents (*Agrostis*) and fescues (*Festuca*) (NVC *Festuca rubra* subcommunity of the *Arrhenatheretum.*) Most of these are too young to show much development. Where *F. rubra* forms a dense sward it is difficult for even the usual species of amenity grassland to colonize. If left unmown or only cut annually, then oat-grass (*Arrhenatherum elatius*) and cocksfoot (*Dactylis glomerata*) invade alone with the big umbellifers, yarrow (*Achillea millefolium*) and docks (*Rumex* spp.) and the vegetation rapidly becomes coarse grassland.

Gardens occupy a significant proportion of many urban areas (Davis, 1982). They can harbour native plant species, although animals are usually more welcome, at least in the form of birds, amphibians, butterflies and bees. Teagle (1978) recorded 90 native and naturalized plant species in suburban front gardens in the West Midlands, and over 170 on allotments. Most were common species usually regarded as 'weeds', but with the increasing interest in 'wild gardening' (e.g. Baines, 1985) more uncommon natives are appearing. In time, this may provide a valuable reservoir for popular species that are now rare in rural areas, such as cowslip and prim-

rose, in the same way as garden ponds have become major habitats for frogs and toads. 'Weeds' (native ones in particular) share a number of biological characteristics (Baker & Stebbins, 1965), but different soils types do have distinctive collections (Salisbury, 1961). Thus, small nettle (*Urtica urens*) and hoary pepperwort (*Cardaria draba*) abound in the chalky gardens of Ramsgate, Kent, but are almost unknown on clay and sandstone.

Unusual substrates and their vegetation

Few urban areas are without some form of industrial dereliction, if only abandoned quarries, claypits or railways. The towns of northern England and other industrial areas may have large areas accumulated over two centuries. Much has been reclaimed over the last 30 years, but considerable amounts remain, colonized by some peculiar floras and providing good examples of primary succession. There is a considerable literature on the physical and chemical characters of waste materials (Bridges, this volume). Brief resumés of their colonization are given below: further details can be found in Brierley (1956), Down (1973), Gemmell (1977, 1982), Greenwood & Gemmell (1978), Ash (1983). The range of wastes in urban areas is considerable, but from the aspect of plant colonization they may conveniently be grouped according to pH (Fig. 9.2).

Acidic wastes are epitomized by colliery shales and their relations such as cinders and clinker. The pH varies from 3 (or less) to 5 or even 6, depending on the strata and coalfield involved. Similar communities may develop on acidic quarry waste such as sandstone. When new, the wastes are severely deficient in nitrogen and phosphorus. They have poor physical structure, are subject to erosion, drought and temperature extremes, and the more recent ones are compacted by heavy machinery. The first colonizers are not annuals, presumably because the nutrient supply is too poor to allow reproduction in one season. Instead, they are perennial grasses (mainly bent, *Agrostis capillaris*, with some Yorkshire fog *Holcus lanatus*) and herbs (especially hawkweed *Hieracium sabaudum*, rosebay willowherb *Chamerion angustifolium* and coltsfoot *Tussilago farfara*). Birch and willow seedlings (*Betula pendula*, *Salix cinerea*, *S. caprea*) may colonize early, especially if seed sources are close at hand, but grow extremely slowly (Curtis, 1977). In most

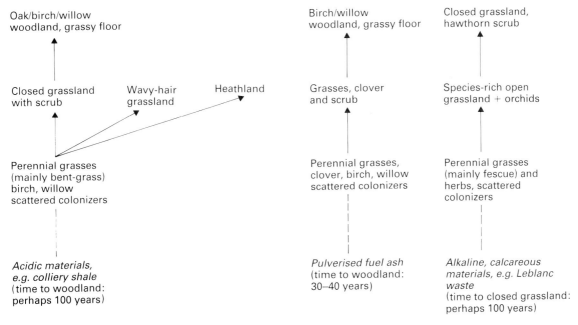

Fig. 9.2 Succession on industrial wastes in north-west England.

cases the vegetation slowly closes over, taking 50 years or more depending on the hostility of the waste, and becomes grassland with scrub. This gradually matures to birch/willow woodland. If oaks are available, and suitable carriers for their acorns (jays, people), they will colonize at the grassland stage giving oak (*Quercus robur*)/birch/willow woodland after some 100 years. Oak gradually predominates as the closing ground cover and shade prevent regeneration of the other species. Eventually it may be difficult to detect the colliery heap beneath the semi-natural vegetation.

Sometimes the succession proceeds instead to heathland, with heather (*Calluna vulgaris*), crowberry (*Empetrum nigrum*) and wavy hair-grass (*Deschampsia flexuosa*), or to acidic grassland dominated by wavy hair-grass, matgrass (*Nardus stricta*) and/or purple moor-grass (*Molinia caerulea*). Availability of seed is obviously one factor directing the succession: heather will grow on colliery shale colonized by bent-grass if seeded artificially (Ash, 1983), but other factors, chemical and physical, remain unexplored. Acidic quarry materials follow similar paths, but may sometimes develop gorse scrub (*Ulex europaeus*), an uncommon species on colliery sites.

Alkaline wastes are a less common feature of urban areas, and calcareous quarries even rarer. Examples are blast furnace slag from iron-smelting and, in north-west England, Leblanc waste from the nineteenth century manufacture of sodium carbonate. When raw, both may have pH of 8–10, but older wastes weather to be similar to limestone soils (Ash, 1983). Nitrogen and phosphorus are very deficient, the latter subject to high adsorption, and there are problems of drought, erosion, temperature extremes, and, on blast furnace slag, a very stony texture. Again most annuals and lower plants are unable to colonize; possibly the nutrient supply is too poor to allow reproduction in one season. The succession starts with perennials tolerant of low nutrient levels such as red fescue (*Festuca rubra*), cocksfoot (*Dactylis glomerata*), knapweed (*Centaurea nigra*) and coltsfoot (*Tussilago farfara*). As the grassland develops, arrays of marsh orchids (*Dactylorhiza* spp. and hybrids) and fragrant orchid (*Gymnadenia conopsea*) often appear and a species-rich community can develop, with carnation-grass (*Carex flacca*), yellow-wort (*Blackstonia perfoliata*), mouse-ear hawkweed (*Hieracium pilosella*), bird's-foot trefoil (*Lotus corniculatus*) and many other species, including rare and peculiar ones. One large Leblanc waste heap, Nob End near Bolton, is now owned by the local authority and managed as a country park because of its flora (Greenwood & Gemmell, 1978; Smyth, 1987). It has recently been declared an SSSI. Such vegetation is a strange mixture, composed of the relatively few local species that can tolerate the severe edaphic problems, plus a range of calcicoles which have managed to cross the considerable distances from 'natural' sources, accidental introductions of various sorts (blue-eyed grass, *Sisyrinchium bermudiana*, grows on two sites) and a few species outside their normal ecological range. Fiorin (*Agrostis stolonifera*) and hemp agrimony (*Eupatorium cannabinum*), both marsh species, grow on Leblanc waste in St Helens.

These calcareous wastes are some of the few habitats where birch and willow do *not* colonize, perhaps because they are so prone to drought. The only shrub to appear is usually hawthorn (*Crataegus monogyna*), which grows very slowly: century-old sites still have little scrub growth and are easily eroded to bare ground by trampling. The vegetation remains open for decades; limestone grassland species were successfully introduced to 50-year-old deposits of Leblanc waste and blast furnace slag without displacing the existing species (Ash, 1983).

Pulverized fuel ash (PFA) is possibly the commonest modern industrial waste to be deposited in urban areas. It is produced by burning pulverized coal in power stations. Compared to many wastes it is fairly benign, having a silty texture, usually adequate supplies of phosphorus and potassium, and a pH about 8 when first tipped. However nitrogen is deficient, the waste often forms cemented layers impenetrable to plant roots, and most fresh deposits contain considerable amounts of boron, which is toxic to plants. The latter leaches to acceptable levels over five to ten years (less if the ash is lagooned), but the colonizing species tend to be those tolerant of high boron levels.

Some sites are initially saline and early colonizers may include hastate orache (*Atriplex prostrata*) and red goosefoot (*Chenopodium rubrum*), species which naturally grow in the upper reaches of salt marshes. More frequently the colonizers are perennial grasses and herbs such as Yorkshire fog (*Holcus lanatus*), red fescue (*Festuca rubra*), cocksfoot (*Tussilago farara*), horsetail (*Equisetum arvense*) and clovers (*Trifolium repens, T. pratense*) as well as the inevitable marsh orchids, along with willows and

birch. The clovers are at a double advantage, being boron-tolerant as well as fixing their own nitrogen, and can dominate a site in its early years. Succession proceeds to willow/birch woodland. It is not known whether other species will eventually invade.

The extreme edaphic conditions of these wastes present two problems for colonizers: species must not only be able to live on such sites, they first have to reach them. Species from limestone grassland and calcareous dunes were spectacularly successful in establishing on Leblanc waste and blast furnace slag in north-west England when sown there. They had no way of colonizing naturally from source areas 30–40 km away. The only calcicole species which had crossed the distance were those with mobile, usually wind-dispersed, seed, such as the orchids. In consequence 'waste heap' floras are more impoverished than they need to be, and offer ideal opportunities for 'creative conservation' (Ash, 1983; Bradshaw, 1986).

Railways, active or closed, are ubiquitous in urban areas. The vegetation of active ones was recently the subject of a national study (Sargent, 1984). Unfortunately, this was virtually confined to rural areas, but some features appear to be true also of urban tracks. The track bed itself, if used, supports only such few species as can avoid spraying with herbicides or recolonize rapidly afterwards. The verge vegetation depends to some extent on soils (often modified by tipping of spent ballast on flat or embanked stretches), but coarse grassland dominated by false-oat (*Arrhenatherum elatius*) was by far the commonest community. Next in abundance was bramble, followed by scrub woodland of very varied composition. Little distinctive 'railway vegetation' was discerned, although the high frequency of rosebay willow-herb (*Chamerion angustifolium*) may be connected with burning in the past as much as with its ability to colonize spent ballast. Once closed and the ballast removed, railway tracks gradually colonize from the edges. The compacted central strip is difficult for roots to penetrate and often becomes a footpath, horsetrack or motor-bike track – official or unofficial.

In contrast to PFA or railways, some wastes are confined to only a few areas, for historical reasons. Thus, the Leblanc process was restricted to north-west England: Widnes, St Helens and a scattering of sites around Bolton and Bury. St Helens has a unique collection of wastes from the glass industry, the most important ecologically being the 'Burgy Banks', made of a mixture of sand and jeweller's rouge, formerly used to polish plate glass. The waste was deposited behind bunds, rising to 15 m in height and covering several hectares. Initially alkaline, it is largely dominated by tall fescue (*Festuca arundinacea*) and boasts a population of broomrape (*Orobanche minor*), which is locally uncommon. Water leaching from the base of the banks is highly alkaline (pH > 10) and supports several small 'salt-marshes' of sea poa (*Puccinellia distans*). One heap, topped with sand washings and war-time sandbags as well as Burgy waste, displays a variety of vegetation from remnants of marsh through sand dune species and other calcicoles, to willow scrub with good bird populations, and is of major ecological importance to the town.

Other industrial towns have different examples, often of interest for the industrial archaeology and history of the town as well as for their ecology. Indeed, in many old industrial centres (St Helens, Wigan and the West Midlands are notable examples), many sites of great interest and value to wildlife are found on 'derelict land' (Davis, 1976; Greenwood & Gemmell, 1978; Teagle, 1978; Cole, 1986; Ash, 1988). A botanical survey of St Helens Metropolitan Borough was carried out in 1981–1983 for Operation Groundwork (the St Helens, Knowsley and Sefton Groundwork Trust). Out of 38 top-class sites, 20 were on derelict land, plus 25 out of 59 sites of lower, but still significant, quality (St Helens WAG, 1986). These urban sites are particularly important as the surrounding countryside is intensively farmed, with little space for wildlife and poor access for education or recreation. The potential of derelict land for these uses has only recently been recognized. It necessitates a major rethink in reclamation techniques (and public funding) away from destruction and creation of a formal landscape, and towards enhancement, naturalistic treatments and interpretation.

Environmental influences

TRAMPLING

People make footpaths where they want to go, which is not necessarily where surfaced footpaths

run. Consequently, all urban vegetation types may get informal footpaths trodden through them. Grasslands tend to suffer most, marshes and dense scrub least.

The first effects of trampling are to increase certain perennial species which can tolerate such stress. These include rosette species such as daisy (*Bellis perennis*) and plantain (*Plantago major*) and species which root at stem nodes, e.g. white clover (*Trifolium repens*) (NVC *Mesotrophic Grasslands*). Grasses can still maintain a good cover at this stage, since they grow from basal meristems. In amenity areas, ryegrass (*Lolium perenne*) and meadow-grass (*Poa pratensis*) persist, whereas in other habitats these species may invade, helped on poor soils by the eutrophication which generally accompanies footpaths (see below). With more severe trampling pressure, bare patches appear at least periodically, e.g. after winter poaching or summer drought. These are colonized by annuals such as *Poa annua*, the introduced pineapple weed (*Matricaria matricoides*), knotweed (*Polygonum aviculare*) and pearlworts (*Sagina* spp.) (Shimwell, 1983). As trampling pressure increases further, the perennials die out to leave only the annuals, frequently growing in wheel-ruts, potholes and other sheltered micro-habitats. The final stage is bare earth, with an edge of less trampled vegetation, varying in width depending on the soil type, drainage and surrounding vegetation.

This pattern appears to apply over most urban soils, but where nutrients are very deficient, as on industrial wastes, the annuals cannot grow. Here light use is usually sufficient to take the vegetation directly from the daisy/clover stage to bare ground.

EUTROPHICATION

Because many urban soils are deficient in nitrogen, and sometimes other nutrients also, local eutrophication can have noticeable effects on the vegetation. Small eutrophic patches, such as those due to dogs or foxes, may merely cause increased grass growth. This is particularly noticeable on many reclaimed sites, especially those with a low legume component in the sward. Trees frequently have a ring of high N and P round them, which may be beneficial, especially where their trunks are kept clear of competing vegetation using herbicide sprays (Bradshaw, 1982, 1986). Particularly rich spots, usually disturbed, may be marked by stands of wall barley (*Hordeum murinum*).

Where larger areas of eutrophication occur (e.g. banks of polluted streams, edges of rubbish dumps, where garden refuse is fly-tipped), perennials of rich soils establish. These include stinging nettle (*Urtica dioica*), docks (*Rumex* spp.), thistles (*Cirsium* spp.) and smaller species such as goosegrass (*Galium aparine*) and deadnettles (*Lamium purpureum, L. album*). This vegetation is attractive to wildlife, but is not looked on with favour by the general public.

Over all urban areas, nitrogen deposition is at least $30 \, kg \, ha^{-1} \, year^{-1}$ – probably more when dry and wet deposition are taken into account (Bradshaw, 1982). This must affect urban vegetation, especially as N deficiency is the limiting factor in many urban ecosystems, pushing plant communities towards those typical of fertile ground. It also renders the preservation of oligotrophic systems, including many attractive communities, more difficult, whether these are water bodies, encapsulated countryside or newly-created habitats.

AIR POLLUTION

Thirty years ago most urban centres had no lichens, and poor bryophyte floras, because of air pollution, in particular the high levels of sulphur dioxide and dust. Since then conditions have improved, although the effects of car exhausts and central heating have partially replaced those of coal smoke. A survey of lichens and the tarspot fungus (*Rhytisma acerinum*) in Merseyside was published in 1980 (Alexander & Henderson-Sellers), building on an earlier one of 1973–1974 (Vick & Bevan, 1974). The sector of central Liverpool which was devoid of lichens in the earlier survey had by the later one been colonized by the most resistant lichen species (the indicator species used was *Lecanora conizaeoides*). This corresponded to a mean winter level of sulphur dioxide of about $170 \, \mu g \, m^{-3}$. Zones of air pollution were mapped using other indicator species: *Xanthoria parietina* being taken as equivalent to $125 \, \mu g \, m^{-3}$ SO_2, tarspot fungus $85–90 \, \mu g \, m^{-3}$, *Parmelia saxatilis* $50–60 \, \mu g \, m^{-3}$. Between the two surveys the inner limits of both *Xanthoria* and tarspot moved considerably towards the centres of Liverpool, Birkenhead and other towns. The fungus (an annual growing on sycamore leaves) moved faster than the perennial lichens, which have very slow growth rates and often poor dispersal abilities. The same pattern could be found in other urban centres, with recolonization proceeding steadily if slowly. The rates of spread were very variable and appeared to be affected by availability of substrates, water supply and temperature extremes as well as pollution (Seaward, 1982).

Higher plants may also be affected by air pollution, though data are limited. The common species of urban areas (grasses, clovers, willow, ash) are very tolerant of sulphur dioxide, ozone and nitrous oxides (Saunders & Ward, 1974). Needle-leaved conifers are much more sensitive to SO_2 (Numata, 1982; Grodzinska, 1982). Examples of damage to trees by urban and industrial emissions are frequent (e.g. Grodzinska, 1982; Handley, 1986). De la Chevallerie (1986) questioned whether modern city trees will reach even the size of nineteenth-century plantings, since high SO_2 has been to some extent replaced by NO_x and the trees are subject to other stresses including soil compaction and poor soil aeration. 'Acid rain' has also had its effects: farmers in Everton in the first decade of the nineteenth century were warring with a certain Mr Muspratt over the acid emissions from his Leblanc works (Hardie, 1950). Most of the older, inner city, Liverpool parks have acidic grassland (e.g. with large amounts of bent-grass *Agrostis capillaris*), where soils over sandstone, naturally inclined to acidity, have been worsened by air pollution (Handley, 1986). Such swards are poor and rapidly break up under heavy use. There is evidence that the moderate levels of SO_2 now prevailing in urban areas of the United Kingdom can inhibit the growth of grass species, and also that populations from polluted sites possess greater tolerance to SO_2 pollution than populations from clean sites. Such tolerance can arise in commercial cultivars by natural selection in only four to five years, when sown in urban conditions (Bell *et al.*, 1982).

CLIMATE

Hollis (this volume) mentions the differences in climate between urban and rural areas. The greater rainfall in towns is offset by high runoff and high temperatures, leading to reduced available water and increased evapo-transpiration. Urban vegetation frequently suffers from drought, especially in dry climates (Miess, 1979). This can be exacerbated by poor water retention in some disturbed soils, and frequently by compaction or high stone content which restrict rooting depth. Light intensity is frequently reduced by clouds and dust. However, carbon dioxide concentrations may be large, raised by combustion of fossil fuels and retained by the relatively low wind speeds, particularly in large conurbations.

The effects of these factors and pollution on general plant growth have not been thoroughly investigated. Possibly most species grow less and die earlier in urban areas (Capel, 1980; Wittig & Durwen, 1982; de la Chevallerie, 1986). The small water supply, interacting with poor soils, explains the high failure rate of standard trees in most urban planting schemes. Results could be improved simply by watering, (Dutton & Bradshaw, 1982), although using smaller stock and species which are known to thrive in urban conditions would also be sensible (Insley & Buckley, 1986).

However, the warmth and low humidity in cities encourages various alien plants to establish which do not tolerate the cool damp of the British countryside. Thus fig trees occur as far north as Sheffield and Liverpool; they are abundant (and large) beside the River Avon in Bristol (Gilbert, 1983). Some species reach their northern limit in cities. In Sheffield yellow vetchling (*Lathyrus aphaca*), sickle medick (*Medicago falcata*) and an annual grass (*Vulpia ambigua*) flourish – all are species believed to be native only in dry pastures in southern Britain. In Berlin, one of the major colonizers of brick rubble is an alien Mediterranean species of *Chenopodium*, whose success is attributed to city warmth as well as soil conditions (Henke & Sukopp, 1986). The 'heat island' effect also favours false locust (*Robinia pseudo-acacia*) as a major component of the shrub stage on brick rubble.

Similarities and differences between urban areas

There is a general air of 'sameness' about most of our towns and cities, especially the housing areas, which is only partly due to building styles. The 'greenspace' of British towns largely consists of two communities, amenity and rough grasslands, whose

dominant species are tolerant of a wide range of soils and whose character is largely determined by management (or the lack of it). Amenity grassland is preferred, because it is 'tidy', but neither community is of great interest to people (except for functional areas such as sports pitches), or to wildlife (Teagle, 1978; Clouston, 1986). There are increasing pressures, economic and social, for change. Keeping the amenity grassland mown has assumed priority in the workload of local authorities, dictating the size of the workforce and causing peaks of work in the spring and late summer (Clouston, 1986; Parker, 1986). Faced with ever-tighter budgets, many pieces of land have ceased to be mown, creating more rough grassland, which is regarded as socially undesirable. At the same time interest in nature conservation (urban wildlife in particular), and in countryside recreation have greatly increased, combined with the realization that the rural area no longer has vast reservoirs of wildlife and many habitats have been decimated (Mabey, 1980; Baines, 1986; Goode & Smart, 1986). A recent unpublished survey carried out by Operation Groundwork revealed many projects aiming to solve the problems of turning amenity or rough grassland into something more interesting, from flowery meadows to urban woodland. In a few years time there may be some answers, and perhaps urban grasslands will take on a more varied look, and may even be more closely related to soils.

In contrast to this similarity, the colonizers of untended urban sites, especially on poorer patches of soil, differ considerably from place to place. Bristol is not only noted for figs, it has large populations of *Buddleia davidii*, introduced in 1890 from China, often associated with the native traveller's joy (*Clematis vitalba*). Both are calcicoles, and other lime-loving species have moved in from the limestone cliffs of the Avon Gorge onto brick rubble sites. Seed availability is obviously an important factor affecting the vegetation composition. Similarly in Liverpool, evening primroses (*Oenothera* spp.) are common, spreading in from the sand dunes of the south Lancashire coast, where they have been established since the late nineteenth century. In the northern cities buddleia is a rarity, possibly for lack of seed source. Urban scrub in Newcastle-upon-Tyne consists largely of crab apple (*Malus sylvestris*), elder (*Sambucus nigra*), hawthorn (*Crataegus monogyna*) and bramble (*Rubus futicosus* agg.). Younger sites in the same city often support the calcicoles wild parsnip (*Pastinaca sativa*) and musk thistle (*Carduus nutans*), which are not common in the surrounding area. As with industrial wastes, both soil conditions and propagule supply are obviously important.

Manchester has many wetland species, reflecting the region's climate, such as reed canary-grass (*Phalaris arundinacea*), amphibious bistort (*Polygonum amphibium*) and creeping yellow-cress (*Rorippa palustris*). Leeds (Gilbert, 1983) had large amounts of mugwort (*Artemisia vulgaris*) and wormwood (*A. absinthium*), Birmingham abundant N. American goldenrod (*Solidago canadensis*), and Sheffield a rich mix of garden escapes including feverfew (*Tanacetum parthenium*), tansy (*T. vulgare*), lupin (*Lupinus polyphyllus*), soapwort (*Saponaria officinalis*) and Michaelmas daisy (*Aster novibelgii*). The abundance of many of these lack ecological explanations. Plainly seed availability is only part of the story, albeit an important one, and much ecological work could be done on these urban 'wastelands'.

In addition to the diversity of such volunteer vegetation, there are the opportunities presented by industrial dereliction with extreme soils, and the variety of encapsulated countryside, all of which 'help to preserve the individual character of a town' (Teagle, 1978). They represent valuable links with an area's history, as well as ecology, and could be exploited to develop a town's sense of pride and place, often much diminished in the last few decades (St Helens WAG, 1986). If the 'urban commons' (to use Gilbert's phrase) can be seen as a potential asset, and treated to enhance their 'capabilities', much benefit to people and wildlife could result.

Three factors appear to be vital in determining urban vegetation; soil conditions, propagule supply and management (or its lack). However, urban botany is still a new science; urban zoology lags even further behind. Much work is now being done, and this review could doubtless be rewritten in ten years' time with much greater understanding. We might even have some idea what happened to that most ubiquitous of urban plants, the rosebay willow-herb:

> '*Epilobium angustifolium* is an uncommon native of moist banks and wood-margins. It is valuable in shrubberies, as thriving under the drip of trees, and succeeds everywhere, even in the smoke of cities, and in parks. It is a good plant to adorn pieces of water, being hardy, of rapid increase, not much relished by cattle, and very showy when in flower'.

(Buxton (1849)
Flora of Manchester)

References

Alexander, R.W. & Henderson-Sellers, A. (1980). *Survey of lichens and tar-spot fungus* (Rhytisma acerinum) *as indicators of sulphur dioxide pollution in the county of Merseyside*. Merseyside County Council, Liverpool. Unpublished.

Ash, H.J. (1983). *The Natural Colonisation of Derelict Industrial Land*. PhD thesis, University of Liverpool.

Ash, H.J. (1988). St Helens. *Urban Wildlife* **1**, 14–18.

Baines, J.C. (1985). *How to Make a Wildlife Garden*. Elm Tree Books, London.

Baines, J.C. (1986). Design considerations at establishment. In Bradshaw, A.D., Goode, D.A. & Thorp, E. (eds), *Ecology and Design in Landscape*, pp. 73–81. Blackwell Scientific Publications, Oxford.

Baker, H.G. & Stebbins, G.L. (1965). *Genetics of Colonising Species*. Academic Press, London.

Bell, J.N.B., Ayazloo, M. & Wilson, G.B. (1982). Selection for sulphur dioxide tolerance in grass populations in polluted areas. In Bornkamm, R., Lee, J.A. & Seaward, M.R.D. (eds), *Urban Ecology*, pp. 171–180. Blackwell Scientific Publications, Oxford.

Bradshaw, A.D. (1982). The biology of land reclamation in urban areas. In Bornkamm, R., Lee, J.A. & Seaward, M.R.D. (eds), *Urban Ecology*, pp. 293–303. Blackwell Scientific Publications, Oxford.

Bradshaw, A.D. (1986). Ecological principles in landscape. In Bradshaw, A.D., Goode, D.A. & Thorp, E. (eds), *Ecology and Design in Landscape*, pp. 15–36. Blackwell Scientific Publications, Oxford.

Bradshaw, A.D. & Chadwick, M.J. (1980). *The Restoration of Land*. Blackwell Scientific Publications, Oxford.

Brierley, J.K. (1956). Some preliminary observations on the ecology of pit heaps. *Journal of Ecology* **44**, 383–390.

Capel, J.A. (1980). *The Establishment and Growth of Trees in Urban and Industrial Areas*. PhD thesis, University of Liverpool.

Clouston, J.B. (1986). Landscape design: user requirements. In Bradshaw, A.D., Goode, D.A. & Thorp, E. (eds), *Ecology and Design in Landscape*, pp. 5–14. Blackwell Scientific Publications, Oxford.

Cole, L. (1986). Urban opportunities for a more natural approach. In Bradshaw, A.D., Goode, D.A. & Thorp,

E. (eds), *Ecology and Design in Landscape*, pp. 417–426. Blackwell Scientific Publications, Oxford.

Curtis, M. (1977). *Trees on Tips*. MSc Dissertation, Salford University.

Dancer, W.S., Handley, J.F. & Bradshaw, A.D. (1977). Nitrogen accumulation in kaolin mining wastes in Cornwall. II. Forage legumes. *Plant and Soil* **48**, 303–314.

Davis, B.N.K. (1976). Wildlife, urbanisation and industry. *Biological Conservation* **10**, 249–291.

Davis, B.N.K. (1982). Habitat diversity and invertebrates in urban areas. In Bornkamm, R., Lee, J.A. & Seaward, M.R.D. (eds), *Urban Ecology*, pp. 49–63. Blackwell Scientific Publications, Oxford.

De la Chevallerie, H. (1986). The ecology and preservation of street trees. In Bradshaw, A.D., Goode, D.A. & Thorp, E. (eds), *Ecology and Design in Landscape*, pp. 383–397. Blackwell Scientific Publications, Oxford.

Down, C.G. (1973). Life form succession in plant communities on colliery waste tips. *Environmental Pollution* **5**, 19–23.

Dutton, R. & Bradshaw, A.D. (1982). *Land Reclamation in Cities*. Department of the Environment, London.

Eaton, J.W. (1986). Waterplant ecology in landscape design. In Bradshaw, A.D., Goode, D.A. & Thorp, E. (eds), *Ecology and Design in Landscape*, pp. 285–306. Blackwell Scientific Publications, Oxford.

Gemmell, R.P. (1977). *Colonisation of Industrial Wasteland* Studies in Biology no. 80. Arnold, London.

Gemmell, R.P. (1982). The origin and importance of industrial habitats. In Bornkamm, R., Lee, J.A. & Seaward, M.R.D. (eds), *Urban Ecology*, pp. 33–39. Blackwell Scientific Publications, Oxford.

Gemmell, R.P. (1988). Tip top treasures. *Natural World* **23**, 8–10.

Gilbert, O. (1983). The wildlife of Britain's wasteland. *New Scientist* 24 March, 824–829.

Goode, D.A. & Smart, P.J. (1986). Designing for wildlife. In Bradshaw, A.D., Goode, D.A. & Thorp, E. (eds), *Ecology and Design in Landscape*, pp. 219–235. Blackwell Scientific Publications, Oxford.

Green, B.H. (1986). Controlling ecosystems for amenity. In Bradshaw, A.D., Goode, D.A. & Thorp, E. (eds), *Ecology and Design in Landscape*, pp. 195–209. Blackwell Scientific Publications, Oxford.

Greenwood, E.F. & Gemmell, R.P. (1978). Derelict land as a habitat for rare plants in S. Lancs. (vc59) and W. Lancs. (vc60). *Watsonia* **12**, 33–40.

Grime, J.P. (1979). *Plant Strategies and Vegetation Processes*. Wiley, London.

Grodzinska, K. (1982). Plant contamination caused by urban and industrial emissions in the region of Cracow (Southern Poland). In Bornkamm, R., Lee, J.A. & Seaward, M.R.D. (eds), *Urban Ecology*, pp. 149–160. Blackwell Scientific Publications, Oxford.

Handley, J.F. (1983). Nature in the urban environment. In Grove, A.B. & Cresswell, R.W. (eds), *City Landscapes*, pp. 47–59. Butterworth, London.

Handley, J.F. (1986). Landscape under stress. In Bradshaw, A.D., Goode, D.A. & Thorp, E. (eds), *Ecology and Design in Landscape*, pp. 361–382. Blackwell Scientific Publications, Oxford.

Hardie, D.W.F. (1950). *A History of the Chemical Industry in Widnes*. Imperial Chemical Industries, Runcorn.

Henke, H. & Sukopp, H. (1986). A natural approach in cities. In Bradshaw, A.D., Goode, D.A. & Thorp, E. (eds), *Ecology and Design in Landscape*, pp. 307–324. Blackwell Scientific Publications, Oxford.

Insley, H. & Buckley, G.P. (1986). Causes and prevention of establishment failure in amenity trees. In Bradshaw, A.D., Goode, D.A. & Thorp, E. (eds), *Ecology and Design in Landscape*, pp. 127–141. Blackwell Scientific Publications, Oxford.

Kunick, W. (1982). Comparison of the flora of some cities of the Central European Lowlands. In Bornkamm, R., Lee, J.A. & Seaward, M.R.D. (eds), *Urban Ecology*, pp. 13–22. Blackwell Scientific Publications, Oxford.

Laurie, I.C. (1979). Urban commons. In Laurie, I.C. (ed.), *Nature in Cities*, pp. 231–266. Wiley, London.

London Ecology Unit (1988). *London's Meadows and Pastures*. LEU Handbook no. 8. London Ecology Unit, London.

Mabey, R. (1980). *The Common Ground*. Hutchinson, London.

Miess, M. (1979). The climate of cities. In Laurie, I.C. (ed.), *Nature in Cities*, pp. 91–114. Wiley, London.

Nature Conservancy Council (no date). *Phase One Habitat Mapping Manual*. Nature Conservancy Council, Peterborough.

National Vegetation Classification (no date). *Mesotrophic Grasslands*. c/o J. Rodwell, Lancaster University.

Numata, M. (1982). Changes in ecosystem structure and function in Tokyo. In Bornkamm, R., Lee, J.A. & Seaward, M.R.D. (eds), *Urban Ecology*, pp. 139–148. Blackwell Scientific Publications, Oxford.

Parker, J.C. (1986). Low cost systems of management. In Bradshaw, A.D., Goode, D.A. & Thorp, E. (eds), *Ecology and Design in Landscape*, pp. 211–218. Blackwell Scientific Publications, Oxford.

Peterken, G.F. (1986). Ecological elements in landscape history. In Bradshaw, A.D., Goode, D.A. & Thorp, E. (eds), *Ecology and Design in Landscape*, pp. 55–68. Blackwell Scientific Publications, Oxford.

Rackham, O. (1980). *Ancient Woodland: Its History, Vegetation and Uses in England*. Arnold, London.

St Helens Wildlife Advisory Group (1986). *A Policy for Nature* (draft). Available from: The Land Manager, Community Leisure Department, Century House, Hardshaw Street, St Helens, Merseyside.

Salisbury, E.J. (1961). *Weeds and Aliens*. Collins, London.

Sargent, C. (1984). *Britain's Railway Vegetation*. Institute of Terrestrial Ecology, Cambridge.

Saunders, P.J.W. & Ward, C.M. (1974). Plants and air pollution. *Landscape Design* **104**, 28–30.

Seaward, M.R.D. (1982). Lichen ecology of changing urban environments. In Bornkamm, R., Lee, J.A. & Seaward, M.R.D. (eds), *Urban Ecology*, pp. 181–189. Blackwell Scientific Publications, Oxford.

Scott, D., Greenwood, R.D., Moffatt, J.D. & Tregay, R.J. (1986). Warrington New Town: an ecological approach to landscape design and management. In Bradshaw, A.D., Goode, D.A. & Thorp, E. (eds), *Ecology and Design in Landscape*, pp. 143–160. Blackwell Scientific Publications, Oxford.

Shimwell, D.W. (1983). *A Conspectus of Urban Vegetation Types*. Unpublished. School of Geography, The University, Manchester M13 9PL.

Smyth, B. (1987). *City Wildspace*. Hilary Shipman, London.

Sukopp, H., Blume, H. & Kunick, W. (1979). The soil, flora and vegetation of Berlin's waste lands. In Laurie, I.C. (ed.), *Nature in Cities*, pp. 115–132. Wiley, London.

Teagle, W.G. (1978). *The Endless Village*. Nature Conservancy Council, Shrewsbury.

Tregay, R. & Moffatt, D. (1980). An ecological approach to landscape and management at Oakwood, Warrington. *Landscape Design* **132**, 33–36.

Vick & Bevan, R.J. (1974). Report on the biological methods used in the detailed mapping of air pollution in Merseyside County. Merseyside County Council, Liverpool. Unpublished.

Wittig, R. & Durwen, K. (1982). Ecological indicator-value spectra of spontaneous urban floras. In Bornkamm, R., Lee, J.A. & Seaward, M.R.D. (eds), *Urban Ecology*, pp. 23–31. Blackwell Scientific Publications, Oxford.

Woodell, S.R.J. (1979). The flora of walls and pavings. In Laurie, I.C. (ed.), *Nature in Cities*, pp. 135–157. Wiley, London.

Plant names are cited according to Clapham, A.R., Tutin, T.G. & Warburg, E.F. (1981) *Excursion Flora of the British Isles*, 3rd Edition. Cambridge University Press, Cambridge.

Appendix
A key to the major urban vegetation types

Shimwell: A Conspectus of Urban Vegetation Types (1983)

1 Pioneer, open communities of rocks, walls, roofs, pavements and other metalled or trodden surfaces, dominated by cryptogams or other specially-adapted rock plants (lithophytes).

1a Carpets of acrocarpous (upright and tufted) mosses and flat thalloid liverworts; flowering plants absent or sparse and of low vitality.

1b Lichen-dominated communities of asbestos roofs and concrete.

1c Miscellaneous fragmentary stands of ferns and flowering plants in crevices and on ledges.

2 Floating and submerged aquatic communities composed of perennial, obligate hydrophytes of fresh waters.

2a Communities of free-floating, surface and submerged plant species (pleustophytes) of static or slow-moving waters.

2b Communities of floating and submerged aquatic species (rhizophytes) of static or slow-moving waters.

2c Bryophyte communities of periodically or permanently submerged rocks and walls of rivers and canals.

3 Ruderal weed communities dominated by annual plant species (therophytes) or biennial species.

3a Open communities of low-growing annuals of gardens, ornamental park borders, roadsides and refuse tips.

3b Therophyte-dominated communities of derelict brick-rubble, cinder and fuel-ash tips, etc. (includes communities with large numbers or even mostly perennials).

3c Communities of rich eutrophic muds and silts around reservoirs, sewage beds and periodically inundated sites.

4 Emergent tall, swamp communities of rivers, canals and lake margins dominated by a variety of linear-leaved monocotyledons.

4a Reedbeds dominated by *Phragmites*.

4b Grass and bulrush-swamp communities.

4c Tall sedgemarsh communities dominated by medium-sized (70–150 cm), rhizomatous species of *Carex*.

5 Low-growing (usually < 70 cm) swamps and marsh communities, dominated by graminoid and herbaceous perennial marsh plant species (helphytes).

5a Miscellaneous low-growing, emergent, species-poor swamps.

5b Rush marshes dominated by species of *Juncus*.

5c Sedge marshes dominated by species of *Carex*.

6 Rank, perennial, tall-grass and tall-herb (usually >70 cm) communities of embankments, road verges, abandoned sewage beds and damp swamp and marsh margins.

6a Communities dominated by tall, coarse grasses and umbellifers.

6b Communities dominated by tall, gregarious, native herbs.

6c Communities dominated by introduced plant species, many of garden origin; (this group also includes communities dominated by the prolific annual, *Impatiens glandulifera*).

7 Low, perennial grass and grass-herb communities, including regularly mown grasslands.

7a Grasslands on a variety of *natural habitats and soils*, dominated by one or several low-growing (<70 cm) fine-leaved, grass species.

7b Managed, mown grasslands and weedy, perennial herb-grass communities of mainly man-made land, in which *rye-grass and/or white clover* predominate.

7c Communities dominated by either introduced or native low-growing (<70 cm), gregarious, stoloniferous or rhizomatous *herbs*.

8 Scrub vegetation of thickets, hedges and ornamental park borders, dominated by woody shrubs less than 5 m in height.

8a Dwarf scrub dominated by ericaceous (heath-like) species less than 70 cm tall.

8b Gorse and broom thickets of acidic soils.

8c Bramble patches in a variety of habitats.

8d Mixed woodland-edge scrub and hedgerows dominated by hawthorn, elder, hazel and a variety of other native shrub species, on dry or moist soils.

8e Birch, willow, alder and poplar scrub of damp soils, either naturally developed or planted.

8f Scrub of introduced, evergreen and deciduous shrubs, either in managed ornamental situations or naturalized in waste places.

9 Deciduous and evergreen woodland, greater than 5 m in height with a ± closed canopy, and urban savannah woodland with an open canopy.

9a Oak woodlands on relatively dry, acidic soils.

9b Beech plantation woodlands.

9c Mixed or pure sycamore, elm, ash plantation woodlands, copses and coverts on a variety of soil types.

9d Willow, alder and birch woodlands of damp, peaty or silty soils with a high water table.

9e Ornamental plantation woodlands and avenues in which the predominant trees are introduced deciduous species, hybrids or cultivars.

9f Coniferous plantations of both deciduous and evergreen species.

9g Managed, urban savanna woodland, comprising expanses of mown grass (see 7b) with clumps of native or introduced, standard trees.

Index